Sub-Contracts DOM/1 and DOM/2

A Guide to Rights and Obligations

Don Riding
ARICS

b

Blackwell Science

© 1996 by
Blackwell Science Ltd
Editorial Offices:
Osney Mead, Oxford OX2 0EL
25 John Street, London WC1N 2BL
23 Ainslie Place, Edinburgh EH3 6AJ
238 Main Street, Cambridge
  Massachusetts 02142, USA
54 University Street, Carlton
  Victoria 3053, Australia

Other Editorial Offices:
Arnette Blackwell SA
  224, Boulevard Saint Germain
  75007 Paris, France

Blackwell Wissenschafts-Verlag GmbH
  Kurfürstendamm 57
  10707 Berlin, Germany

  Zehetnergasse 6
  A-1140 Wien
  Austria

First published 1996

Set in 11 on 13pt Palatino
by DP Photosetting, Aylesbury, Bucks
Printed and bound in Great Britain by
Hartnolls Ltd, Bodmin, Cornwall

DISTRIBUTORS

Marston Book Services Ltd
PO Box 269
Abingdon
Oxon OX14 4YN
(*Orders:* Tel: 01235 465500
         Fax: 01235 465555)

USA
Blackwell Science, Inc.
238 Main Street
Cambridge, MA 02142
(*Orders:* Tel: 800 215-1000
         617 876-7000
         Fax: 617 492-5263)

Canada
Copp Clark, Ltd
2775 Matheson Blvd East
Mississauga, Ontario
Canada, L4W 4P7
(*Orders:* Tel: 800 263-4374
         905 238-6074)

Australia
Blackwell Science Pty Ltd
54 University Street
Carlton, Victoria 3053
(*Orders:* Tel: 03 9347 0300
         Fax: 03 9347 5001)

A catalogue record for this title is available
from the British Library

ISBN 0-632-04125-0

Library of Congress
Cataloging-in-Publication Data
Riding, Don.
    Sub-contracts DOM/1 and DOM/2:
  a guide to rights and obligations/
  Don Riding.
      p.   cm.
    Includes index.
    ISBN 0-632-04125-0
  1. Construction contracts—Great Britain.
  2. Construction industry—Subcontracting
  —Great Britain.   I. Title.
  KD1641.Z9R53   1996
  343.41'078624—dc20
  [344.10378624]
                            96-26491
                                 CIP

Contents

Preface

The author of this book has spent the last thirty or so years in the construction industry, most of that time as a quantity surveyor for a major national building contractor. During that period he has been acutely aware that sub-contractors, irrespective of the size or complexity of the organisation, rarely understand and appreciate their rights and obligations under the standard forms of sub-contract.

Many books are available covering the interpretation and application of the main building contracts in common use (JCT 80, GC/Works/1 etc.) and these include guidance on how to submit claims under those contracts, but few publications have been produced with sub-contractors and sub-contracts in mind.

Analysis shows that upwards of 60 per cent of construction works is sub-let and the sub-contractor, therefore, plays a major role within the industry. It is also a sad fact that, with the construction industry hard hit by the recession, there has been a considerable decrease in profitability which has in turn seen an increase in problems of payment and dispute. It is more than ever essential, therefore, that sub-contractors should fully understand their contractual commitments. Indeed, a better understanding of those commitments should lead to a reduction in disputes, prompter payment and increased profitability.

It would be possible to cover both nominated and domestic sub-contracts, nominated being where the architect has reserved to himself the final selection and approval of the sub-contractor to supply and fix any materials or goods or execute work, but nominated sub-contractors have become less and less popular over recent years, mainly as a result of a perception that the employer and/or architect assume a greater responsibility for such sub-contractors. Where an architect still wishes to retain control over the choice of which sub-contractor to use, there is now a tendency to specify or allow selection from a named list and by far the majority of sub-contracts, therefore, are now let on a domestic basis with the contractor assuming full responsibility for the actions of the sub-contractor.

Hence, this book which deals with the rights and obligations of sub-contractors under the Sub-Contracts DOM/1 and DOM/2. There are almost three hundred individual rights and/or obligations in these sub-contracts and it would be naive to expect sub-contractors to be aware of them all. Most will be familiar with the broad principles of delay, extensions of time and loss and expense, but not the fine detail in connection with those matters nor the broader aspects concerned with, for example, insurance, payment or practical completion.

The text on Sub-Contract DOM/1, for use where the main contract is the JCT Standard Form of Building Contract (1980), takes account of Amendments 1–9 of DOM/1 and that on Sub-Contract DOM/2, for use where the main contract is the JCT Standard Form 'With Contractor's Design' (1981), of Amendments 1–7 of DOM/2. It would be possible to reproduce in full the sub-contracts, but that would merely fill out the contents of this book and add to the cost of production. The chosen format, therefore, is to produce a list of clause/condition numbers with any relevant headings and side headings only, the main contract clause headings being shown in bold type in a slightly enlarged font and side clause headings in bold but in the normal text size. Against each clause/condition number is a statement as to the sub-contractor's relevant right and/or obligation followed by any other comment considered relevant. All clause numbers are included for completeness, but not all clauses contain a right or obligation or warrant further comment. The text is not intended to cover every situation likely to arise in a contract; it is intended to contain sound professional advice in general terms covering the prime objectives of the book, namely that sub-contractors should fully understand their rights and obligations.

This book represents the views of one quantity surveyor on DOM/1 and DOM/2 – it is not intended to be a definitive legal interpretation and in any particular circumstance readers should obtain independent advice on the law and its applications to their particular facts.

The impact of the Housing Grants, Construction and Regeneration Act 1996 on DOM/1 and DOM/2 is very limited, the Act being restricted to resolution of contentious issues, i.e. adjudication, payment and contra charges. The clauses affected are:

(a) *Adjudication:* The idea of adjudication to resolve disputes prior to instituting arbitration proceedings is a totally new concept. At present in DOM/1 and DOM/2, adjudication is only available under clause 24 to resolve contractors' claims not agreed by the sub-contractor. It is an interim measure only prior to arbitration and notice for arbitration has to be given at the same time as institution of the procedure.

(b) *Payment:* As written, clause 21 of DOM/1 and DOM/2 does not contain 'pay-when-paid' provisions, but many contractors amend the standard wording to include such provisions. Those amendments will be against the requirements of the Act and, therefore, unenforceable in the future.

(c) *Set-off:* The current set-off provisions in clause 23 comply with the requirements of the Act.

(d) *Suspension for non-payment:* Such a right is already embodied within clause 21.6 of DOM/1 and DOM/2.

D Riding
August 1996

Part 1

Basic Principles

The Function of a Contract

What is a contract?

Before starting to look in detail at the parties' rights and obligations under the standard forms of sub-contract, it is necessary to understand what a contract is.

A contract is a promise which the law will enforce, although there are contracts where one party has so behaved as to lead the other to believe that he has given a promise and, in those circumstances, he will be precluded from denying the existence of a contract.

Most contracts therefore have:

(1) An offer or conditional promise by one party.
(2) An acceptance of that offer by the other party.
(3) An intention on both sides to create a legal obligation and not merely to perform a social function, still less to perform an illegal act.
(4) There must be consideration which was defined by the courts as long ago as 1875 as 'some right, interest, profit or benefit accruing to one party or some forbearance, detriment, loss or responsibility given, suffered or undertaken by the other'. However, that is less than comprehensive since it does not make clear that consideration may be a promise given in exchange for a promise and there is consideration if the promisor (or some third party at his request) does benefit even if it costs the promisee nothing.
(5) There must be no duress or undue influence.
(6) The contract must be made by persons professing capacity to contract.

There are two aspects to the final statement above:

(1) There is nothing to stop a person who is not authorised to bind his employer from entering into a contract. In most companies

only certain employees have authority to commit the company to enter into contract; e.g. the receptionist would probably not be empowered to sign a contract and if any other party accepted his/her signature, that contract in law would be against the company but not against the signatory.

(2) As a practical consideration, although not a legal requirement, the person must be capable of honouring his part of the contract; e.g. there is little point in offering a plastering contract to, say, a plumber since the plumber would quite clearly be unable to honour his commitment. There is nothing, however, which prevents a person entering into a contract he cannot perform; for breach of it he will be liable for damages.

The basic requirements above apply to all contracts irrespective of their simplicity or complexity.

How are these requirements incorporated into sub-contracts?

The following statements look at how basic requirements apply to sub-contracts and they apply to all sub-contracts irrespective of the form used.

Generally sub-contractors will be invited to tender for work by main contractors (or consultants acting on their behalf) and the enquiry document(s) should set out clearly all of the parameters which the sub-contractor is to consider in arriving at his price. At this stage the sub-contractors invited to tender will have been selected on the basis of their ability to carry out the various work packages.

In the event of a sub-contractor being successful, he will normally be offered a sub-contract based on one of the standard forms. Note here that no contract exists – there has been no acceptance of any offer. Only on subsequent acceptance of the offered sub-contract by the sub-contractor does a legal binding sub-contract exist, but it is important to note that acceptance can take place in one of two ways:

(1) by formal signing of the contract document(s), or
(2) by action – if a sub-contract has been offered but not actually signed and a sub-contractor acts by performing the sub-contract, he will be deemed to have accepted the sub-contract

by his action, the terms and provisions being as laid down in the offered sub-contract. It is no use acting and later trying to object to onerous or any terms.

What can often happen is that the sub-contractor makes an offer, the main contractor counter-offers and the sub-contractor then accepts, although there may well be several counter-offers by and to the sub-contractor in between times. This process creates a legal obligation and consideration (i.e. agreement to pay) and there is no duress. Thus the basic requirements for a valid contract are satisfied. However it can sometimes be difficult to establish what exactly are the conditions of the contract.

What part does the enquiry play in contract?

Reference has been made to enquiry documents above and these documents will normally be incorporated into the sub-contract as one (or more) of the listed documents. *They are very important* as they set out all of the parameters to be used in arriving at the sub-contract price; get this part wrong and a dispute is inevitable!

The information included with the enquiry should contain (as a minimum):

(1) the form of the head contract, with any amendments thereto, together with the main contract start and completion dates (if known) or the main contract period for completion of the works

(2) the proposed form of sub-contract which should be appropriate for use with the head contract

(3) any supplementary conditions applicable

(4) a list of documents enclosed with the enquiry (detailed page numbers, drawing numbers etc.). If a list is not provided, the sub-contractor should carefully record all of the documents and set them out in his tender to clearly establish the basis of his pricing (particularly as greater emphasis is now placed upon drawn information issued with the enquiry following the introduction of co-ordinated project information). Documents issued should include preliminaries, trade preambles/ specification and bills of quantities (where applicable); the preliminaries should include main contract conditions and

any amendments thereto together with details of any sectional completion requirements.

If all of the information listed above is not available, the sub-contractor should request such further particulars as are necessary to enable him/her to submit a properly considered bid. Sub-contractors should also be aware that not all main contractors will invite tenders on the same basis; inevitably if a standard form is used it will be amended somewhere by most major contractors. Some do not use one of the standard forms available but invite tenders based on their own form, which will usually contain much more onerous terms and conditions, and it is, therefore, important that sub-contractors fully consider the various enquiries and reflect the more onerous conditions in their pricing.

If, as often happens, a sub-contractor receives several enquiries for the same contract from different main contractors and submits the same price to them all irrespective of the particular conditions proposed by each, there is no pressure on the more unscrupulous main contractors to change their methods and adopt one of the standard forms or remove more onerous conditions.

Equally it is not sufficient to qualify bids where onerous conditions or unacceptable sub-contracts are contemplated; they will invariably be ignored when the sub-contract is offered. The only way to change bad practice is by pricing the more onerous conditions and giving different bids to different contractors. In this way the more reasonable contractors will receive more realistic bids and the more unscrupulous may eventually go out of business or at least win less work. Indeed there is merit in providing two bids – one a price for using a standard form and an extra over price for the particular proposed conditions; in this way the more unscrupulous main contractors will come to realise what their unreasonable conditions are costing.

One must, however, have regard to the fact that more and more onerous conditions are being placed upon main contractors by employers and, in these circumstances, stepping down of that risk is inevitable. One should also bear in mind that qualifications in a bid can lead to disqualification of that bid in total, particularly if the bid is sought by a consultant, and qualification should, therefore, only be considered where the proposed conditions are either totally unacceptable or cannot be priced.

Sub-contractors should also be aware that 'Standard Conditions of Sale' forming part of a bid are equally likely to be ignored on sub-contract placement or can conflict with terms within the contract itself and are, therefore, to a large extent a pointless waste of time. If one of the standard forms is proposed nobody should have any objections since all parties involved in the construction industry were consulted during its drafting.

So there are the basic principles to be aware of before entering into contract and in the following chapters the rights and obligations of sub-contractors under the Domestic Sub-Contracts DOM/1 and DOM/2 will be considered in detail.

Part 2

Domestic Sub-Contract DOM/1

Introduction

The Domestic Sub-Contract DOM/1, which is published by the Building Employers Confederation, is intended for use where the form of main contract is the JCT 80 Standard Form of Building Contract: Local Authorities/Private edition With/Without Quantities (JCT 80), and it has been approved by the Building Employers Confederation, the Specialist Engineering Contractors Group, the Federation of Associations of Specialists and Sub-Contractors and the Federation of Building Specialist Contractors.

A total of nine amendments have been issued since the first edition in 1980, Amendment 1 (1984), 2 and 3 (1987), 5, 6 and 8 (1989) and 9 (1990) being now incorporated as standard. Amendment 4 to Sub-Contract DOM/1 (published in September 1989 and relating to Supplementary Provisions incorporating Schedules of Work and Contract Sum Analysis) is only for use where Amendment 3 to JCT 80 applies to the main contract and is applicable only to the Without Quantities version; Amendment 7 to Sub-Contract DOM/1 (also published in September 1989 and relating to the Standard Method of Measurement 7th edition) is only for use where Amendment 7 to JCT 80 applies to the main contract. These two Amendments, 4 and 7, will be considered further at the end of Part 2.

The Sub-Contract DOM/1 comprises two documents: the articles of agreement and the sub-contract conditions. The next chapter will consider the articles, moving on to the conditions later. It is, however, important to understand that *the conditions by themselves do not constitute a sub-contract – a sub-contract can only exist where the articles of agreement have been completed.*

The Articles of Agreement

The articles of agreement actually comprise two parts, the first being the agreement appertaining to each individual contract and the second part being the appendix applicable to that individual contract.

The articles open with provision for inserting the date and the names and registered addresses of the parties to the contract, i.e. 'the Contractor' and 'the Sub-Contractor'. There then follow four recitals or statements covering:

1. The contractor desires to have work executed as referred to in the appendix.
2. The sub-contract works are to be carried out as part of work carried out by the contractor under a contract with the employer and provision is made in the second recital to insert the name of the employer.
3. The sub-contractor has had reasonable opportunity of inspecting all of the provisions of the main contract.

Note here that the sub-contractor has a right to inspect 'all of the provisions of the Main Contract, or a copy thereof, except the detailed prices of the Contractor included in schedules and bills of quantities' and the sub-contractor must be given a reasonable opportunity of carrying out that inspection.

The emphasis given to 'all' is the author's and the sub-contractor should ensure that he does so inspect before entering into contract. Any later claims of 'not being aware' will undoubtedly fail because of the statement in this recital and the later article 1.1. Note also particular comments regarding enquiry documents in Part 1 of this book.

4. The relevant tax status of sub-contractor, contractor and employer under the provisions of the Finance (No.2) Act 1975

and which of clauses 20A or 20B is to apply under the sub-contract.

The recitals are followed by three articles which set out what is agreed:

1.1. The sub-contractor is deemed to have notice of all of the provisions of the main contract. Whether the sub-contractor has, in fact, notice of all of the provisions is immaterial – he is 'deemed to have notice' and hence the importance of the note under recital 3 above.

1.2. The sub-contractor shall carry out and complete the sub-contract works.

This article contains the overlying obligation of the sub-contractor 'to carry out and complete the Sub-Contract Works', but subject to the sub-contract documents *and the provisions of the main contract*. One cannot ignore what is in the main contract and rely solely on what is in the sub-contract, although in the event of conflict, and conflict only, the terms of the sub-contract documents prevail (clause 2.2 of the conditions).

1.3. The sub-contract conditions are deemed to be incorporated. Provision is made within this article to incorporate any amendments which are to apply.

2. Article 2 allows the insertion of the sub-contract sum which is to be inclusive of any cash discount. There are two sub-articles 2.1 and 2.2 and one is to be deleted as appropriate, 2.2 being for use where the sub-contract works are to be completely re-measured and valued (footnote [c]).

The completion of article 2.1 does not, however, mean that no alteration to quantities can take place. In the event of changes brought about by drawing issues and/or architect's instructions, or in the event of an error in quantity where the main contract is a With Quantities form, they would be variations to be measured and valued in accordance with the rules laid down in the sub-contract conditions.

3. In the event of a dispute or difference, except a dispute or difference under clause 20A (Finance (No.2) Act 1975 – Tax Deduction Scheme) or clause 20B (Finance (No.2) Act 1975 – Tax Deduction Scheme – Sub-Contractor not user of a current tax certificate) to the extent provided in clause 20A.8 or 20B.6 (both relating to where the act or the regulations or any other act of parliament or statutory instrument, rule or order made under an act of parliament provide for some other method of resolving the dispute), both parties agree that such dispute or difference shall be resolved by arbitration in accordance with clause 38.

Finally in the first part of the articles of agreement there is provision for formal signing and execution either under hand or as a deed.

The appendix to DOM/1, which then follows, contains a series of statements regarding both the main contract and sub-contract which, together with the recitals and articles already mentioned, individualise the Sub-Contract DOM/1 such that it becomes a unique document applicable only to the contract contained therein. The appendix itself does not impose any additional rights or obligations beyond those envisaged in the sub-contract.

Part 1   Section A is referred to in the second recital and provides for completion of details of the main contract in respect of:

   (1)   the works, which is to be the same description as in the main contract articles of agreement (footnote [e])
   (2)   the form of main contract conditions
   (3)   where the main contract documents may be inspected
   (4)   whether the main contract is executed as a simple contract or as a deed
   (5)   alternative provisions within the main contract, and
   (6)   any amendments to or from the printed standard form identified.

Section B provides for completion of details of the main contract appendix and entries therein.

Section C provides for completion of details in respect of obligations or restrictions imposed by the employer and not covered by the main contract conditions, any employer's requirements affecting the order of the works and the location and type of access.

   Having regard to comments made earlier regarding enquiry documents and inspection of main contract documents, all of the above should contain nothing new and should merely be confirmation of what the sub-contractor should already know.

Parts 2 to 14 then contain details applicable to the sub-contract only as follows:

Part 2   provides for completion of details in respect of particulars of the sub-contract works and the numbered documents to be annexed to Sub-Contract DOM/1. Note that the Sub-Contract DOM/1 comprises the articles of agreement *and* the conditions and it is not, therefore, necessary to list them here. Only those documents additional to DOM/1, e.g. enquiry, tender, any supplemental conditions etc., need be listed.

Part 3   provides for completion of details in respect of the amount of insurance cover the sub-contractor is required to maintain to cover his liability for any injury or damage to property as required by clause 7.2 of the conditions.

Part 4   provides for completion of details in respect of carrying out and completion of the sub-contract works. The details required to be completed are the dates between which the sub-contract works will be commenced, the period required for carrying out and completion of those works on site, the period required for notice to commence work on site, the period required (if any) for sub-contract works off site and prior to commencement on site and, finally, any further details.
Note that start and completion dates are not entered. The trigger for the sub-contract as regards carrying out and completion is the notice to commence, but no notice is required for works off site and it is the sub-contractor's responsibility to so arrange matters that any off site work is completed in time to commence on site. Equally, sufficient time for off-site works must be allowed by the contractor before giving notice to commence on site.

Part 5   provides for completion of details in respect of rates or prices and/or daywork rates or prices to be used in the valuation of variations under clause 16 or the valuation of all work comprising the sub-contract works under clause 17 of the conditions.

Part 6 provides for completion of details in respect of which of the alternative VAT provisions is to apply.

Part 7 provides for completion of details in respect of cash discounts and retention percentages. Note that the cash discount is 2½% *unless a different percentage is inserted*; if this section is left blank, payments are subject to a cash discount of 2½%, not nil as some believe.

Part 8 provides for completion of details in respect of the adjudicator and trustee-stakeholder referred to in clause 24 of the conditions.

Part 9 provides for completion of details in respect of any particular items of attendance to be provided other than those detailed in clause 27 of the conditions.

Part 10 provides for completion of details in respect of which of the fluctuations clauses is to apply.

Clause 35 is for use where the sub-contract is on a 'fixed price' basis except for changes in the types and rates of contribution, levy and tax payable by a person in his capacity as an employer. Reimbursement is made on the basis of actual changes against the list of items included in part 11 of the appendix.

Clause 36 is for use where the sub-contract is fully fluctuating and reimbursement is made on the basis of actual prices against tender prices for the list of items included in part 12 of the appendix.

Clause 37 is for use where reimbursement is made on a formula basis.

In the event of a 'fully fixed price' sub-contract, clauses 35, 36 and 37 will be deleted.

Part 11 provides for completion of details in respect of the list of items to be used for reimbursement of fluctuations under clause 35 of the conditions.

Part 12 provides for completion of details in respect of the list of items to be used for reimbursement of fluctuations under clause 36 of the conditions.

Part 13 provides for completion of details in respect of the formula rules to be applied under clause 37 of the conditions.

Part 14 provides for completion of details in respect of the appointment of the arbitrator in the event of a dispute and arbitration under clause 38.

Part 15 provides for completion of details in respect of any other matters agreed between the contractor and sub-contractor, such as special conditions or agreements on employment of labour, limitation on working hours and the like.

The Conditions

1 Interpretation, definitions etc.

1.1

1.2

1.3

2 Sub-contract documents

2.1 Sub-contract documents – documents other than sub-contract documents

2.2 Relationship of sub-contract documents

3 Sub-contract sum – additions or deductions – computation of ascertained final sub-contract sum

Clause 3 clearly establishes the sub-contractor's right to payment even if the item is only partially ascertained and agreed, and partial in this sense is exactly what it says and can, for example, refer to quantity without agreement on price or vice versa.

The author is aware of main contractors who suggest that unless an item is agreed, no payment will be made against that item. Equally the author is aware of cases where no payment is made on the grounds that the main contractor has not been paid under the main contract. Do not be so coerced – both of these requirements contravene the requirements of clause 3.

Payment under this clause is not linked to payment under any other contract, nor does the sub-contract contain 'pay when paid' provisions and the sub-contractor has a right to payment irrespective of the main contract situation.

The only exception to this statement is the date of final payment (clause 21.9.2) which is dependent upon the issue of the final certificate under clause 30.8 of the main contract conditions, but dependent upon issue only – it is not dependent upon payment having been made.

4 Execution of the sub-contract works – directions of contractor

4.1 Sub-contractor's obligations to carry out and complete the sub-contract works

4.1   .1
This is affirmed in article 1.2, the articles setting out clearly the particular contract to which these general conditions apply.

The obligation here is to carry out and complete the sub-contract works not only in compliance with the sub-contract documents but in conformity with all reasonable directions and requirements of the contractor.

The directions or requirements of the contractor are only in so far as they may apply to regulating for the time being the due carrying out of the works; the sub-contractor is not under any obligation to conform with all directions or requirements, only those for which he has a specific obligation under the sub-contract conditions.

4.1   .2, .3, .4
These three sub-clauses all contain the word 'shall', which places an obligation on the sub-contractor to ensure that all materials, goods and workmanship comply with the standards.

Failure to comply with these requirements can have disastrous effects on the financial outcome of a project, since the sub-contract confers upon the contractor considerable powers in the event of breaches of these requirements, and sub-contractors should, therefore, ensure, by way of quality management systems or the like, that everything, as far as possible, complies with the documents. It is not unknown for contractors and sub-contractors to suggest that standards vary depending upon the nature of the particular contract and that prestigious, high quality offices would be built to a higher standard than, say, local authority housing. The finishings to the office may be more expensive, but the standards to which the materials, goods or workmanship comply are exactly the same if the same relevant British Standards apply to both.

Where approval of quality or standards is a matter for the opinion of the architect, such quality and standards shall be to the reasonable satisfaction of the architect.

This is the second time that the word 'reasonable' has been used and it is not the last. The word appears several times in DOM/1 which is unfortunate, since what is reasonable is a matter of opinion and can be a recipe for conflict. In this context the statement only refers to those items of materials, goods and workmanship which are not described in the sub-contract documents as being to British Standard or equivalent or to some other standard, e.g. DIN Standards, and where they are described as being 'to the satisfaction of the Architect'. As a general rule all materials specified will be covered by some standard or other and this statement is included within the sub-contract to cover the remote possibility of some material being selected by the architect which cannot comply with a relevant published standard or for which no relevant standard has been published. It is suggested that, in interpretation of the word reasonable in this context, custom and practice within the relevant trade would have to be considered.

4.1   .5

Discrepancies in or divergences between documents

If any discrepancy or divergence is found the sub-contractor is obliged to give to the contractor a written notice specifying the discrepancy or divergence.

This obligation, however, only exists *if* any discrepancy or divergence is found; the sub-contractor is not under any obligation to find such discrepancies or divergences.

In the event that the written notice is given, the contractor is obliged to issue directions in regard thereto and the sub-contractor has a right, per se, to receive those directions and, it is suggested, to receive those directions timeously such that there is no effect upon the sub-contractor's ability to carry out and complete the sub-contract works within the period specified in the sub-contract.

4.2 Directions of contractor
.1, .2, .3, .4

The sub-contractor has an obligation to comply with any direction issued in accordance with clause 4.2. However, the sub-contractor has the right not to so comply if the direction is one which is not reasonable or is one requiring a variation within the definition of variation in clause 1.3, paragraph 2, but he then has an obligation to make reasonable objection *in writing* to such compliance.

Objection and non-compliance can only be made under the specific circumstances listed and must be reasonable. So, objection and non-compliance to an ordered variation because 'the price has not been agreed' is not considered to be reasonable grounds, since rules for the valuation of variations are contained in clause 16 or clause 17 as appropriate, and failure to comply under these circumstances could lead to additional costs for which the sub-contractor would be liable.

4.3 .1
Inspection – tests

Unlike clause 4.2, the sub-contractor does not have any right to objection, rather the reverse – he has a specific obligation to open up or test as directed.

4.3 .2
Powers of contractor – work not in accordance with the sub-contract

4.3 .2 .1

Unless the direction is issued pursuant to clause 4.2.2 following the issue of instructions under clause 8.4.1 of the main contract conditions, the sub-contractor has a right of consultation and regard to Code of Practice 'A' appended to the Sub-Contract Conditions DOM/1 (for which see the next chapter) prior to the issue of the said direction, and/or

4.3   .2 .2

In the event that directions are issued, the sub-contractor will be obliged to comply with those directions, and/or

4.3   .2 .3

The sub-contractor has a right of regard to Code of Practice 'B' appended to the Sub-Contract Conditions DOM/1 (for which see later in this chapter).

The contractor may not issue any direction under clause 4.3.2 which relates to materials or goods or workmanship where approval of the quality or standards is a matter for the opinion of the architect unless such direction is issued pursuant to clause 4.2.2.

Where any case of non-complying work is the subject of an instruction issued by the architect pursuant to clause 8.4 of the main contract conditions, the sub-contractor has the right to be so informed.

4.3   .3 .1

This clause relates purely to situations where non-complying work is allowed to remain with the agreement of the employer.

The clause places an obligation on the contractor to consult with the sub-contractor and it follows that the sub-contractor, therefore, has a right to such consultation.

The sub-contractor does not have a right to consultation in all cases of non-complying work but only within the specific situation mentioned.

4.3   .3 .2

This clause places an obligation on the contractor to inform the sub-contractor in writing where non-complying work is allowed to remain and the sub-contractor is obliged to allow an appropriate deduction to be made.

What 'appropriate deduction' means is not specified and will be a matter for negotiation between the parties; a calculated sum for the 'changed' specification is unlikely to be acceptable and an element of 'commercial awareness' will be expected. Indeed the architect is likely to ask for the amount of the deduction before agreeing to the work remaining and sub-contractors should bear in mind the probable rectification costs in arriving at this figure – too low a figure is likely to lead to an instruction under 4.3.2.

4.3   .3   .3

The sub-contractor is obliged to accept and perform all that is required by the architect's instruction and, the instruction arising from non-complying work (a breach of the express provisions of clause 4.1 of the sub-contract), no adjustment shall be made to the sub-contract sum or be included in the computation of the ascertained final sub-contract sum and no extension of time shall be given.

4.3   .3A

Powers of contractor – non-compliance with clause 4.1.4

The contractor does not have to issue directions but, in the event that directions are issued, the sub-contractor is obliged to accept and perform all that is required by those directions and, the direction arising from a breach of the express provisions of clause 4.1.4 of the sub-contract (relating to carrying out work in a proper and workmanlike manner), no adjustment shall be made to the sub-contract sum or be included in the computation of the ascertained final sub-contract sum and no extension of time shall be given. The sub-contractor does, however, have a right to consultation *before* the contractor issues directions under this clause.

Where failure to comply with clause 4.1.4 is the subject of an instruction issued under clause 8.5 of the main contract, the contractor shall so inform the sub-contractor and the sub-contractor, therefore, has the right to be so informed.

Note here that any direction pursuant to clause 8.5 of the

main contract will be issued under clause 4.2.2 of the sub-contract and the sub-contractor has an obligation to comply with any direction issued under clause 4.2 except in so far as the direction is not reasonable or is one requiring a variation within the definition of variation in clause 1.3 (see previous commentary under clause 4.2 for more details).

4.3 .4 .1
The sub-contractor is obliged to take down etc. any work properly executed if so directed by the contractor. If the sub-contractor is the 'offending party', he has the right to be advised *in writing* of what work properly executed by others has to be taken down and re-executed etc.

4.3 .4 .2
The sub-contractor has the right to be paid for directions issued under clause 4.3.4.1, the rules for ascertainment of the value of the payment and when the payment is to be made being as stated in the clause.

The sub-contractor must be aware that if he is the 'non-complying party' then no instructions will be issued under clause 4.3.4.1 and he will not be entitled to payment – rather the reverse. Any costs incurred by others, which may include the contractor, other sub-contractors and suppliers, for taking down and re-executing properly executed work will be passed to the non-complying sub-contractor in accordance with the following clause 4.3.5.

4.3 .5 .1, .2
Note the use of the word 'shall' which places an obligation on the sub-contractor to indemnify and reimburse costs all as stated in the clause.

4.4 Directions otherwise than in writing
The contractor has the right to issue directions under clause 4.2 but they should be in writing.

If they are not in writing then the sub-contractor is obliged to confirm in writing any such direction within 7 days and if not dissented from in writing by the contractor within 7 days from the receipt of the sub-contractor's confirmation any such direction takes effect as from the expiration of the latter 7 days.

There are very few occasions when sub-contractors comply with the confirmation requirements of this clause and the author knows of many incidents where sub-contractors have complied with what they believed they had been instructed only to find much later that their interpretation of the instruction was incorrect. Full compliance with this clause, *by both parties to the sub-contract*, would relieve some of the heartache at final account stage as to what was or was not instructed.

4.4   .1, .2

It is extremely important that all directions are confirmed by one party or the other, not only from the point of view of payment for work properly executed but, and possibly more importantly, to clearly establish the final scope of the sub-contract works and responsibility for defects during and beyond the defects liability period.

4.5   Sub-contractor's failure to comply with directions

The sub-contractor is obliged to comply with all reasonable directions as detailed above but, in the event of non-compliance, this clause places additional obligations on the sub-contractor:

(1) to begin compliance within 7 days as stated in the clause, or

(2) bear the consequence of non-compliance by payment of costs incurred.

All too often sub-contractors believe that they have the 'upper hand' when things go wrong – and it is naive to think that things never go wrong – and try to ignore their duty to correct any deficiency by claiming that it is 'alright' or will be 'put right' without any real intention of doing anything. Correction at the time is inevitably the cheapest option, before people become enmeshed in argument as to what is or is not 'reasonable'. All sub-contractors should, therefore, make themselves aware of:

(1) their obligations defined in clause 4.1 of the sub-contract relating to the carrying out and completion of the sub-contract works
(2) the considerable powers of the architect under clause 8 of the main contract relating to inspection, tests and work not in accordance with the contract
(3) the considerable powers of the contractor under clause 4 of the sub-contract relating to inspection, tests, work not in accordance with the sub-contract and non-compliance with clause 4.1.4 (all work to be carried out in a proper and workmanlike manner)

and ensure that, as far as reasonably possible (and this may involve quality assurance and control systems), they comply with their obligations.

5   Sub-contractor's liability under incorporated provisions of the main contract

5.1

Note the wording 'The sub-contractor shall' which places an obligation on the sub-contractor to do whatever is specified under clause 5.1.

5.1   Sub-contractor to observe etc. all provisions of main contract

5.1   .1

The sub-contractor has an obligation to observe, perform and comply with *all* of the provisions of the main contract as referred to in the appendix part 1. This is a general statement confirming article 1.2, which obliges the sub-contractor to carry out and complete the sub-contract works subject to the sub-contract documents and the provisions of the main contract, but, without prejudice to the generality of this statement, the clause further obliges the sub-contractor to specifically observe, perform and comply with the main contract conditions:

clause 6   Statutory obligations, notices, fees and charges
clause 7   Levels and setting out of the works
clause 9   Royalties and patent rights

clause 16   Materials and goods unfixed or off-site
clause 32   Outbreak of hostilities
clause 33   War damage, and
clause 34   Antiquities

5.1   .2   .1, .2, .3

These sub-clauses are preceded by the opening words of clause 5.1, 'The Sub-Contractor shall', and the sub-contractor is obliged, therefore, to comply with them all.

Note particularly that the opening words of clause 5.1.2 require the sub-contractor to 'indemnify and save harmless the Contractor against and from' the items listed and these words should not be taken lightly. The dictionary definition of indemnify is 'to protect or insure against penalties incurred by actions; to compensate for injury suffered; to reimburse; to give security against' and the sub-contractor is, therefore, agreeing to reimburse all costs arising from any of the matters listed in clause 5.1.2. Failure, therefore, to observe, perform and comply with those listed matters could result in the sub-contractor having to bear considerable costs.

5.2   Acts or omissions of the employer, contractor etc. – exclusion of sub-contractor's liability

6   Injury to persons and property – indemnity to contractor

6.1   Definitions

6.2   Liability of sub-contractor – personal injury or death – indemnity to contractor

This clause obliges the sub-contractor to take responsibility for personal injury to or death of any person except as specifically listed in the clause and further requires the sub-contractor to indemnify (see comment after clause 5.1.2 above) the contractor against any claims arising therefrom.

This would be effected by way of insurance (see clause 7.1 below).

6.3 Liability of sub-contractor – injury or damage to property – indemnity to contractor

Again this clause obliges the sub-contractor to take responsibility for any injury or damage to any property as listed in the clause and further requires the sub-contractor to indemnify (see comment after clause 5.1.2 above) the contractor against any claims arising therefrom.

This would again be effected by way of insurance (see clause 7.1 below).

6.4 Extent of liability and indemnity under clause 6.3 – exclusion

The 'specified perils' are defined in clause 1.3 of DOM/1 (fire, lightning, explosion, storm, tempest, flood etc.) and the sub-contractor has the right to protection from these matters, whether or not caused by the negligence, breach of statutory duty, omission or default of the sub-contractor, for the period up to and including whichever is the earlier of the terminal dates (as defined in clause 6.1.2).

7 Insurance against injury to persons or property

7.1 Insurance – personal injury or death – injury or damage to property

The sub-contractor is obliged to take out and maintain insurance in respect of claims for personal injury to or the death of any person (as referred to in clause 6.2) or claims for injury or damage to any property real or personal (as referred to in clause 6.3) as modified by clause 6.4, except that his obligation for insurance against injury or damage to property as stated in clause 6.3 shall not extend to taking out and maintaining insurance for injury and damage to the sub-contract works by a risk other than a specified peril up to and including whichever is the earlier of the terminal dates.

It is important for sub-contractors to further read and understand footnote [c] which amplifies the requirements of the insurance required to be taken out under clause 7.1, namely:

The sub-contractor has the benefit of the main contract joint names policy for loss or damage by the specified perils to the sub-contract works but not for other risks e.g. subsidence, impact, theft or vandalism. The insurance to which clause 7.1 refers is a third party or public liability policy; such a policy will not however give cover for any property such as the sub-contract works whilst they are in the custody and control of the sub-contractor. As the sub-contractor is liable for those other risks if they cause loss or damage to the sub-contract works he may well consider that he needs to take out a works insurance to provide such cover as he does not get under the main contract joint names policy, but he is not obliged to do so; he can accept the risk himself, but he is still indemnifying the contractor under clause 6.

The sub-contractor has the right to the benefits of the joint names policy taken out by the contractor under the main contract but this only provides benefit in respect of the specified perils. As the sub-contractor is obliged to indemnify the contractor under clause 6 he should consider carefully if additional insurance is required to enable him to meet his obligations.

It is equally of vital importance that sub-contractors also read and understand footnote [e] which is reproduced verbatim and which requires the sub-contractor to 'consider whether he should take out insurance to cover any risks *for which he is not covered under clause 8A.1*, e.g. impact, subsidence, theft, vandalism. It should also be noted that, since the benefit of the main contract joint names policy is only extended to 'Domestic Sub-Contractors' as defined in clause 19.2 of the main contract conditions, such benefit only extends to the sub-contractor himself and *'not to any company, firm or person to whom the Sub-Contractor sub-lets any portion of the Sub-Contract Works'*. (The emphasis is the author's.)

The benefits of the joint names policy to which the sub-contractor has a right are limited to insurance in respect of the specified perils only, and that benefit extends only to the sub-contractor himself; sub-sub-contractors have no rights to any such benefits.

Many sub-contractors believe that they have no responsibility for the sub-contract works on site and believe that that responsibility rests solely with the contractor, e.g. many believe that if some material or equipment which has been fixed in position is stolen or damaged, the replacement costs are borne by the contractor. The requirements of the various insurance clauses and explanatory footnotes clearly demonstrate that the reverse is the case, except in so far as sub-contract materials or goods have been fully, finally and properly incorporated into the works (for which see clause 8A.2.2, clause 8B.2.2 or clause 8C.2.2 as appropriate).

7.2 Amount of insurance cover

The sub-contractor is obliged to effect insurance under clause 7.1. The obligation under this clause is to ensure that insurance for personal injury complies with the Employer's Liability (Compulsory Insurance) Act 1969, and any relevant statutory orders, amendments or re-enactments thereof. Insurance cover for any other claims is to be not less than the sum stated in the Appendix part 3 for any one occurrence or series of occurrences arising out of one event although the sub-contractor may, if he so wishes, insure for a sum greater than that stated in the Appendix part 3 (footnote [d]).

7.3 Excepted risks

8 Loss or damage to the works and to the sub-contract works

8.1 Sub-contract works and site materials – benefit of joint names policy under main contract – alternative clauses 8A or 8B or 8C

8.2   Exception in clauses 8A.2.1, 8B.2.1 and 8C.2.1

Whilst the sub-contractor has the right to the benefit of the joint names policy under the main contract, he is obliged to take responsibility for the cost of restoration, replacement or repair as set out in clauses 8A.2.1 or 8B.2.1 or 8C.2.1 as applicable if no claim is made under the joint names policy or to the extent that the joint names policy contains excess provisions.

The sub-contractor should ensure that in the event of loss or damage by the specified perils the contractor does indeed make a claim under the joint names policy and should enquire as to excess provisions against that policy prior to tendering so that full and proper consideration of the risks can be made for inclusion in his tender price.

8.3   No modification of sub-contractor's obligations on defects

8.4   Main contract works insurance – payment of insurance monies to employer

8.5   Loss or damage by specified perils – computation of amounts payable to sub-contractor

8A   Sub-contract works in new buildings – main contract conditions clause 22A

Immediate attention is drawn to footnote [e] which requires the sub-contractor to consider whether he should take out insurance to cover any risks *for which he is not covered under clause 8A.1*, e.g. impact, subsidence, theft, vandalism. It should also be noted that, since the benefit of the main contract joint names policy is only extended to 'Domestic Sub-Contractors' as defined in clause 19.2 of the main contract conditions, such benefit only extends to the sub-contractor himself and *not to any company, firm or person to whom the sub-contractor sub-lets any portion of the sub-contract works*.

Many sub-contractors believe that security on site is the main contractor's total responsibility and that should anything, once fixed, go missing or be removed for any reason he has the right to recover the replacement and/or refixing costs from the main contractor. Not so; the terms of the Sub-Contract DOM/1 and particularly footnote [e] clarify that it is the sub-contractor's responsibility to protect his work from any cause, either by way of insurance or by carrying the risk himself and he has, therefore, an obligation to bear that risk. He should also ensure that that risk is stepped down to his own sub-sub-contractors and also the risk arising from the specified perils, since that risk is clearly not borne by the main contractor under the joint names policy in respect of sub-sub-contractors.

8A.1 Benefit of main contract joint names policy for loss or damage by the specified perils

The sub-contractor has the right to recognition under and the benefits of the joint names policy taken out by the contractor under the main contract but only in so far as any loss or damage is caused by the specified perils – see particularly footnote [e] referred to above.

8A.2 Responsibility of sub-contractor – loss or damage to sub-contract works and sub-contract site materials before practical completion of the sub-contract works

8A.2 .1
Note the use of the word 'shall' which places an obligation on the sub-contractor to bear the responsibility for the costs of restoration, replacement or repair except only as specified in the clause, but reference to paragraphs 2 and 3 shows that the exceptions are very limited.

The specified perils are defined in clause 1.3, namely fire, lightning, tempest, flood, bursting or over-flowing of water tanks, apparatus or pipes, earthquake, aircraft and other aerial devices or articles dropped therefrom, riot and civil commotion but excluding excepted risks. Excepted risks are also defined in clause 1.3, namely ionising radiations or contamination by radioactivity from any nuclear fuel or from any nuclear waste from the combustion of nuclear fuel, radioactive toxic explosive or other hazardous properties of any explosive nuclear assembly or nuclear component thereof, pressure waves caused by aircraft or other aerial devices travelling at sonic or supersonic speeds. Thus any loss or damage arising from the specified perils, and this would realistically only arise from the first five listed events (fire, lightning, tempest, flood, bursting or over-flowing of water tanks, apparatus or pipes), is insured by the contractor, and the sub-contractor is covered for any claims under the provisions of clause 8A.1.

All sub-contractors should be aware that should they wish to recover loss or damage arising from any negligence, breach of statutory duty, omission or default of the contractor etc., they will have to be prepared to prove their claim. For example, if some pipework has been fixed and it is stolen or damaged, it is no good the sub-contractor pleading that 'someone has stolen or damaged my pipe-work' and expect to be paid for replacement and/or repair. He will have to produce evidence to show clearly who stole or caused the damage, or produce evidence to show clearly that the contractor or any other person listed in clause 8A.2.1, paragraph 3, was negligent or in breach of statutory duty or omitted to do something or was in default in order for his claim to succeed. In all other cases the sub-contractor will have to bear the costs himself or claim against his own insurance, hence the importance of the sub-contractor clearly understanding his obligations and insurance requirements.

8A.2   **.2**

Note the word 'shall' and the sub-contractor is obliged to take responsibility for the restoration of work lost or damaged but only to the extent that such loss or damage 'is caused by the negligence, breach of statutory duty, omission or default of the Sub-Contractor or any person for whom the Sub-Contractor is responsible' or only to the extent that sub-contract materials or goods have not been 'fully, finally and properly incorporated into the Works'.

'Fully, finally and properly incorporated' has not yet been defined in law and is open to interpretation. It does not generally cause major problems in practice except in respect of plumbing, mechanical or electrical installations and the like where the author has come across interpretations ranging from a bracket fixed to a wall at one extreme to water flowing in a pipe or electricity switched on at the other. The latter interpretation cannot be correct since clause 8A.2.2 refers to sub-contract materials or goods being incorporated into the works 'before practical completion of the Sub-Contract Works'. The wording clearly anticipates the stage of full, final and proper incorporation being reached before practical completion and it is suggested that either of the extremes quoted above is unreasonable. It appears clear that the sub-contract perceives the situation that as the sub-contract works progress, parts of the work become completed and the sub-contractor moves on leaving the completed parts behind to become the responsibility of the contractor. A more realistic interpretation of the meaning of the words, therefore, would be a clearly defined operation in an area, e.g. first fix mechanical on the ground floor, although again when or whether this is complete could be a cause of argument and if it is a large contract, e.g. a hospital, the areas may need to be refined further.

8A.3 **Loss or damage occurring to sub-contract works and sub-contract site materials before practical completion of the sub-contract works – sub-contractor's obligation to restore etc. such loss or damage**

This clause places two obligations on the sub-contractor; firstly to notify the contractor in writing upon discovery of any loss or damage, irrespective of the cause, and secondly to restore the damage and proceed with the sub-contract works.

Sub-contractors should not attempt to hide any loss or damage for which they are responsible nor should they refuse to restore any loss or damage until payment has been agreed: these actions would clearly constitute breaches of the sub-contract for which the sub-contractor would be liable. Equally do not refrain from proceeding with the carrying out and completion of the sub-contract works for such action would also constitute a breach of the sub-contract.

8A.4 Payment for restoration etc. of work done under clause 8A.3 by sub-contractor

This clause gives the sub-contractor the right to payment for loss or damage which is not the responsibility of the sub-contractor.

The valuation for the payment is to be made in accordance with the valuation rules under clause 16 or clause 17 as applicable; the sub-contractor does not have any right to automatic reimbursement on a cost plus basis if the restoration work is similar to that described in the bills of quantities (if quantities apply) nor to reimbursement against a quotation or any other method which departs from the rules laid down in clause 16 or clause 17.

8A.5 Loss or damage occurring to the sub-contract works after their practical completion – responsibility of sub-contractor

The sub-contractor has no responsibility for any loss or damage after practical completion of the sub-contract works (except as stated in the clause) and *has no responsibility to restore any work lost or damaged* unlike clause 8A.3 which requires the sub-contractor to restore lost or damaged work irrespective of responsibility. Should the contractor require lost or damaged work to be restored and requests the sub-contractor to carry out the said restoration work, the sub-contractor need not be bound by the rules laid down in clauses 16 or 17 (see clause 8A.4) since he will have honoured his sub-contract obligations and already have completed his work.

8B Sub-contract works in new buildings – main contract conditions clause 22B

Immediate attention is drawn to footnote [e] which requires the sub-contractor to consider whether he should take out insurance to cover any risks *for which he is not covered under clause 8B.1*, e.g. impact, subsidence, theft, vandalism. It should also be noted that, since the benefit of the main contract joint names policy is only extended to 'Domestic Sub-Contractors' as defined in clause 19.2 of the main contract conditions, such benefit only extends to the sub-contractor himself and *not to any company, firm or person to whom the sub-contractor sub-lets any portion of the sub-contract works.*

Many sub-contractors believe that security on site is the main contractor's total responsibility and that should anything, once fixed, go missing or be removed for any reason he has the right to recover the replacement and/or refixing costs from the main contractor. This is not so. The terms of the Sub-Contract DOM/1 and particularly footnote [e] clarify that it is the sub-contractor's responsibility to protect his work from any cause, either by way of insurance or by carrying the risk himself and he has, therefore, an obligation to bear that risk. He should also ensure that that risk is stepped down to his own sub-sub-contractors and also the risk arising from the specified perils, since that risk is clearly not borne by the main contractor under the joint names policy in respect of sub-sub-contractors.

8B.1 Benefit of main contract joint names policy for loss or damage by the specified perils

The sub-contractor has the right to recognition under and the benefits of the joint names policy taken out by the contractor under the main contract but only in so far as any loss or damage is caused by the specified perils – see particularly footnote [e] referred to above.

8B.2 Responsibility of sub-contractor – loss or damage to sub-contract works and sub-contract site materials before practical completion of the sub-contract works

8B.2 .1
Note the use of the word 'shall' which places an obligation on the sub-contractor to bear the responsibility for the costs of restoration, replacement or repair except only as specified in the clause, but reference to paragraphs 2 and 3 shows that the exceptions are very limited.

The specified perils are defined in clause 1.3, namely fire, lightning, tempest, flood, bursting or over-flowing of water tanks, apparatus or pipes, earthquake, aircraft and other aerial devices or articles dropped therefrom, riot and civil commotion but excluding excepted risks. Excepted risks are also defined in clause 1.3, namely ionising radiations or contamination by radioactivity from any nuclear fuel or from any nuclear waste from the combustion of nuclear fuel, radioactive toxic explosive or other hazardous properties of any explosive nuclear assembly or nuclear component thereof, pressure waves caused by aircraft or other aerial devices travelling at sonic or supersonic speeds. Thus any loss or damage arising from the specified perils, and this would realistically only arise from the first five listed events (fire, lightning, tempest, flood, bursting or over-flowing of water tanks, apparatus or pipes), is insured by the employer, and the sub-contractor is covered for any claims under the provisions of clause 8B.1.

All sub-contractors should be aware that should they wish to recover loss or damage arising from any negligence, breach of statutory duty, omission or default of the contractor etc., they will have to be prepared to prove their claim. For example, if some pipework has been fixed and it is stolen or damaged, it is no good the sub-contractor pleading that 'someone has stolen or damaged my pipework' and expecting to be paid for replacement and/or repair. He will have to produce evidence to show clearly who stole or caused the damage or produce evidence to show clearly that the contractor or any other person listed in clause 8B.2.1, paragraph 3, was negligent or in breach of statutory duty or omitted to do something or was in default in order for his claim to succeed. In all other cases the sub-contractor will have to bear the costs himself or claim against his own insurance, hence the importance of the sub-contractor clearly understanding his obligations and insurance requirements.

8B.2 .2

Note the word 'shall' and the sub-contractor is obliged to take responsibility for the restoration of work lost or damaged but only to the extent that such loss or damage 'is caused by the negligence, breach of statutory duty, omission or default of the Sub-Contractor or any person for whom the Sub-Contractor is responsible' or only to the extent that sub-contract materials or goods have not been 'fully, finally and properly incorporated into the Works'.

'Fully, finally and properly incorporated' has not yet been defined in law and is open to interpretation. It does not generally cause major problems in practice except in respect of plumbing, mechanical or electrical installations and the like where the author has come across interpretations ranging from a bracket fixed to a wall on one extreme to water flowing in a pipe or electricity switched on at the other. The latter interpretation cannot be correct since clause 8B.2.2 refers to sub-contract materials or goods being incorporated into the works 'before practical completion of the Sub-Contract Works'. The wording clearly anticipates the stage of full, final and proper incorporation being reached before practical completion and it is suggested that either of the extremes quoted above is unreasonable. It appears clear that the sub-contract perceives the situation that as the sub-contract works progress, parts of the work become completed and the subcontractor moves on leaving the completed parts behind to become the responsibility of the contractor. A more realistic interpretation of the meaning of the words, therefore, would be a clearly defined operation in an area, e.g. first fix mechanical on the ground floor, although again when or whether this is complete could be a cause of argument and if it is a large contract, e.g. a hospital, the areas may need to be refined further.

8B.3 Loss or damage occurring to sub-contract works and sub-contract site materials before practical completion of the sub-contract works – sub-contractor's obligation to restore etc. such loss or damage

This clause places two obligations on the sub-contractor; firstly to notify the contractor in writing upon discovery of any loss or damage, irrespective of the cause, and secondly to restore the damage and proceed with the sub-contract works.

Sub-contractors should not attempt to hide any loss or damage for which they are responsible nor should they refuse to restore any loss or damage until payment has been agreed: these actions would clearly constitute breaches of the sub-contract for which the sub-contractor would be liable. Equally do not refrain from proceeding with the carrying out and completion of the sub-contract works for such action would also constitute a breach of the sub-contract.

8B.4 Payment for restoration etc. of work done under clause 8B.3 by sub-contractor

This clause gives the sub-contractor the right to payment for loss or damage which is not the responsibility of the sub-contractor.

The valuation for the payment is to be made in accordance with the valuation rules under clause 16 or clause 17 as applicable; the sub-contractor does not have any right to automatic reimbursement on a cost plus basis if the restoration work is similar to that described in the bills of quantities (if quantities apply) nor to reimbursement against a quotation or any other method which departs from the rules laid down in clause 16 or clause 17.

8B.5 Loss or damage occurring to the sub-contract works after their practical completion – responsibility of sub-contractor

The sub-contractor has no responsibility for any loss or damage after practical completion of the sub-contract works (except as stated in the clause) and *has no responsibility to restore any work lost or damaged* unlike clause 8B.3 which requires the sub-contractor to restore lost or damaged work irrespective of responsibility. Should the contractor require lost or damaged work to be restored and requests the sub-contractor to carry out the said restoration work, the sub-contractor need not be bound by the rules laid down in clauses 16 or 17 (see clause 8B.4) since he will have honoured his sub-contract obligations and already have completed his work.

8c Sub-contract works in existing structures – main contract clause 22C

Immediate attention is drawn to footnotes [e] and [f]. Footnote [e] requires the sub-contractor to consider whether he should take out insurance to cover any risks *for which he is not covered under clause 8C.1*, e.g. impact, subsidence, theft, vandalism. It should also be noted that, since the benefit of the main contract joint names policy is only extended to 'Domestic Sub-Contractors' as defined in clause 19.2 of the main contract conditions, such benefit only extends to the sub-contractor himself and *not to any company, firm or person to whom the sub-contractor sub-lets any portion of the sub-contract works.*

Many sub-contractors believe that security on site is the main contractor's total responsibility and that should anything, once fixed, go missing or be removed for any reason he has the right to recover the replacement and/or refixing costs from the main contractor. This is not so. The terms of the Sub-Contract DOM/1 and particularly footnote [e] clarify that it is the sub-contractor's responsibility to protect his work from any cause, either by way of insurance or by carrying the risk himself and he has, therefore, an obligation to bear that risk. He should also ensure that that risk is stepped down to his own sub-sub-contractors and also the risk arising from the specified perils, since that risk is clearly not borne by the main contractor under the joint names policy in respect of sub-sub-contractors.

Footnote [f] confirms that neither clause 8C nor clause 23.3.2 of the main contract conditions provides for the joint names policy referred to in clause 22C.1 of the main contract conditions to be so issued or so endorsed that the sub-contractor is either recognised as an insured under the policy or for the insurers to waive their rights of subrogation.

To the extent that loss or damage is caused by the negligence, breach of statutory duty, omission or default of the sub-contractor or any person for whom the sub-contractor is responsible, he will be liable therefor, such liability being third party liability for which the sub-contractor is expressly liable under clause 6.3 and against which he is obliged to insure under clause 7.1.

Where it is possible that the existing structure/contents could at any time be in the custody or control of the sub-contractor, it is essential that the third party public liability policy should be so worded as to afford adequate protection.

8c.1 Benefit of main contract joint names policy for loss or damage by the specified perils

The sub-contractor has the right to recognition under and the benefits of the joint names policy taken out by the contractor under the main contract but only in so far as any loss or damage is caused by the specified perils – see particularly footnote [e] referred to above.

8c.2 Responsibility of sub-contractor – loss or damage to sub-contract works and sub-contract site materials before practical completion of the sub-contract works

8c.2 .1

Note the use of the word 'shall' which places an obligation on the sub-contractor to bear the responsibility for the costs of restoration, replacement or repair except only as specified in the clause, but reference to paragraphs 2 and 3 shows that the exceptions are very limited.

The specified perils are defined in clause 1.3, namely fire, lightning, tempest, flood, bursting or over-flowing of water tanks, apparatus or pipes, earthquake, aircraft and other aerial devices or articles dropped therefrom, riot and civil commotion but excluding excepted risks. Excepted risks are also defined in clause 1.3, namely ionising radiations or contamination by radioactivity from any nuclear fuel or from any nuclear waste from the combustion of nuclear fuel, radioactive toxic explosive or other hazardous properties of any explosive nuclear assembly or nuclear component thereof, pressure waves caused by aircraft or other aerial devices travelling at sonic or supersonic speeds. Thus any loss or damage arising from the specified perils, and this would realistically only arise from the first five listed events (fire, lightning, tempest, flood, bursting or overflowing of water tanks, apparatus or pipes), is insured by the employer, and the sub-contractor is covered for any claims under the provisions of clause 8C.1.

All sub-contractors should be aware that should they wish to recover loss or damage arising from any negligence, breach of statutory duty, omission or default of the contractor etc., they will have to be prepared to prove their claim. For example, if some pipework has been fixed and it is stolen or damaged, it is no good the sub-contractor pleading that 'someone has stolen or damaged my pipework' and expect to be paid for replacement and/or repair. He will have to produce evidence to show clearly who stole or caused the damage or produce evidence to show clearly that the contractor or any other person listed in clause 8C.2.1, paragraph 3, was negligent or in breach of statutory duty or omitted to do something or was in default in order for his claim to succeed. In all other cases the sub-contractor will have to bear the costs himself or claim against his own insurance, hence the importance of the sub-contractor clearly understanding his obligations and insurance requirements.

8C.2 .2

Note the word 'shall' and the sub-contractor is obliged to take responsibility for the restoration of work lost or damaged but only to the extent that such loss or damage 'is caused by the negligence, breach of statutory duty, omission or default of the Sub-Contractor or any person for whom the Sub-Contractor is responsible' or only to the extent that sub-contract materials or goods have not been 'fully, finally and properly incorporated into the Works'.

'Fully, finally and properly incorporated' has not yet been defined in law and is open to interpretation. It does not generally cause major problems in practice except in respect of plumbing, mechanical or electrical installations and the like where the author has come across interpretations ranging from a bracket fixed to a wall on one extreme to water flowing in a pipe or electricity switched on at the other. The latter interpretation cannot be correct since clause 8C.2.2 refers to sub-contract materials or goods being incorporated into the works 'before practical completion of the Sub-Contract Works'.

The wording clearly anticipates the stage of full, final and proper incorporation being reached before practical completion and it is suggested that either of the extremes quoted above is unreasonable. It appears clear that the sub-contract perceives the situation that as the sub-contract works progress, parts of the work become completed and the sub-contractor moves on leaving the completed parts behind to become the responsibility of the contractor. A more realistic interpretation of the meaning of the words, therefore, would be a clearly defined operation in an area, e.g. first fix mechanical on the ground floor, although again when or whether this is complete could be a cause of argument and if it is a large contract, e.g. a hospital, the areas may need to be refined further.

8C.3 Loss or damage occurring to sub-contract works and sub-contract site materials before practical completion of the sub-contract works – sub-contractor's obligation to restore etc. such loss or damage

8C.3 .1

This clause places an obligation on the sub-contractor to notify the contractor *in writing* upon discovery of any loss or damage as stated.

8C.3 .2

This is a most unlikely occurrence and refers to determination 'if it is just and equitable so to do', but if such an event does happen the provisions of clause 31 apply to the sub-contractor (for which see later).

8C.3 .3

This clause obliges the sub-contractor to restore the damage and proceed with the sub-contract works.

Sub-contractors should not attempt to refuse to restore any

loss or damage until payment has been agreed: this action would clearly constitute a breach of the sub-contract for which the sub-contractor would be liable. Equally do not refrain from proceeding with the carrying out and completion of the sub-contract works for such action would also constitute a breach of the sub-contract.

8c.4 Payment for restoration etc. of work done under clause 8C.3.3 by sub-contractor

This clause gives the sub-contractor the right to payment for loss or damage which is not the responsibility of the sub-contractor.

The valuation for the payment is to be made in accordance with the valuation rules under clause 16 or clause 17 as applicable; the sub-contractor does not have any right to automatic reimbursement on a cost plus basis if the restoration work is similar to that described in the bills of quantities (if quantities apply) nor to reimbursement against a quotation or any other method which departs from the rules laid down in clause 16 or clause 17.

8c.5 Loss or damage occurring to the sub-contract works after their practical completion – responsibility of sub-contractor

The sub-contractor has no responsibility for any loss or damage after practical completion of the sub-contract works (except as stated in the clause) and *has no responsibility to restore any work lost or damaged* unlike clause 8C.3 which requires the sub-contractor to restore lost or damaged work irrespective of responsibility. Should the contractor require lost or damaged work to be restored and requests the sub-contractor to carry out the said restoration work, the sub-contractor need not be bound by the rules laid down in clauses 16 and 17 (see clause 8C.4) since he will have honoured his sub-contract obligations and already have completed his work.

9 Policies of insurance – production – payment of premiums

9.1 Insurance to which clause 7 refers – documentary evidence – policy or policies and premium receipts

Clause 7 obliges the sub-contractor to insure against injury to persons or property and under this clause the sub-contractor is obliged to prove that he has indeed effected and is maintaining the insurance required by clause 7 by producing documentary evidence, *but only as and when reasonably required to do so and subsequently not unreasonably or vexatiously.*

When is 'reasonable' or when is not 'unreasonable or vexatious' are not defined but it is suggested that it would be reasonable to request evidence at sub-contract commencement or placement and subsequently at policy renewal. It would be unreasonable and vexatious to request evidence monthly as a prerequisite to payment, although such a ploy is not unheard of.

9.2 Default by sub-contractor – insurance under clause 7

In the event of default in effecting insurance under clause 7 the sub-contractor is obliged to reimburse the contractor for effecting the necessary insurance *should the contractor elect to insure.*

Note that the contractor is not obliged to take out insurance – clause 9.2 merely states that the contractor 'may' take out insurance, but this does give the contractor the right to insure at the sub-contractor's expense – but if insurance under clause 7 has not been effected it would be a brave contractor who did not exercise his right and arrange the necessary insurance and recover the premium costs as stated.

9.3 Evidence of compliance by contractor with either clause 8A.1 or clause 8B.1 or clause 8C.1

Clause 8A obliges the contractor to insure against injury to persons or property and clauses 8B and 8C oblige the contractor to ensure that such insurance has been effected; this clause is the reverse of clause 9.1 and under this clause the sub-contractor has the right to proof that insurance as required by clause 8A or clause 8B or clause 8C has indeed been effected and is being maintained by producing documentary evidence, but only as and when reasonably required to do so and subsequently not unreasonably or vexatiously and except as stated in the clause.

The reason for the exception is explained in footnote [g], i.e. that the local authorities' version of the standard form does not provide any right for the contractor to require the employer to produce any documentary evidence or the relevant policy or policies and premium receipts.

9.4 Default by contractor – insurance taken out by sub-contractor

In the event of default in complying with clause 9.3 the contractor is obliged to reimburse the sub-contractor for effecting the necessary insurance *should the sub-contractor elect to insure.*

Note that the sub-contractor is not obliged to take out insurance – clause 9.4 merely states that the sub-contractor 'may' take out insurance, but this does give the sub-contractor the right to insure at the contractor's expense – but if it is believed that insurance under clause 8A or clause 8B or clause 8C has not been effected, since clause 9.3 requires documentary proof of such insurance, it would be a brave sub-contractor who did not exercise his right and arrange the necessary insurance and recover the premium costs as stated.

10 Sub-contractor's plant, etc. – responsibility of contractor

The sub-contractor is obliged to assume responsibility for any loss or damage to any temporary works etc. as listed, *unless such loss or damage is due to negligence etc. of the contractor or any person for whom the contractor is responsible.*

Note that the contract only draws specific comment to the contractor or any person for whom the contractor is responsible and is silent as regards loss or damage due to default by persons for whom the employer is responsible (see clause 29 of the main contract conditions); in the absence of specific comment it could be argued that the sub-contractor is also obliged to assume responsibility for such loss or damage.

11 Sub-contractor's obligation – carrying out and completion of sub-contract works – extension of sub-contract time

Immediate attention is drawn to footnote [h] which in turn draws attention to clauses 35.4.7, 36.5.8 and 37.7 (restriction of fluctuations or price adjustment where sub-contractor is in default over completion). Further comment will be made later in this narrative against the relevant clauses quoted.

11.1 Sub-contract works – details in appendix – progress of works

Under this clause the sub-contractor is obliged to carry out and complete the sub-contract works, as already affirmed in article 1.2 (see page 14), but *'in accordance with the details in the Appendix part 4 etc.'*. This clause, therefore, introduces:

(1) the agreed period for completion, and
(2) the requirement for carrying out and completion being reasonably in accordance with the progress of the works

as conditions of the sub-contract.

Each of these items is, however, subject to receipt of the notice to commence work on site (see the appendix part 4) and to the operation of clause 11. Thus the sub-contractor is under no obligation to commence work on site until and unless he has received the notice to commence on site as detailed in the appendix part 4, but once he has received the notice the time clock starts to tick and if the sub-contractor fails to commence in accordance with the notice he will not be entitled to an extension of time and will have to bear the consequences of his breach. Equally the carrying out and completion, once the sub-contractor has received notice, is subject to the operation of clause 11, i.e. he must complete within the period specified in the appendix part 4 unless one of the other sub-clauses in clause 11 allows for variance therefrom.

At this point consideration should be given to what the words 'reasonably in accordance with the progress of the Works' mean and what effect their interpretation will have on completion of the documentation. In the past this phrase has been interpreted by main contractors as a right for them to instruct sub-contractors when and where they will work.

The decision in the case of *Piggott Foundations Limited* v. *Shepherd Construction Limited* (1993) has clarified that no such right is bestowed upon main contractors but rather the reverse, that the sub-contractor has the right to carry out his work how and in whatever order he wishes.

Briefly, the plaintiffs were piling sub-contractors contracted under DOM/1. Delay occurred and the plaintiff sued for the value of its works, applying for summary judgment and for various issues to be determined as points of law. During the hearing, the court was invited to consider a sub-contractor's obligations under clause 11.1 and an official referee sitting in Liverpool held that the sub-contractor was not required to carry out his work in such a manner as either to fit in with any scheme of work of the main contractor or to finish any part of the sub-contract works by a particular date so as to enable the main contractor to proceed with other parts of the works. There was no obligation on the sub-contractor to carry out his work in any particular order or at any specified rate of progress; the sub-contractor was entitled to plan and perform the work as he pleased provided he finished it by the time fixed in the sub-contract.

It is not intended to debate the issues here; what has been decided in law is now the law and the rights and wrongs of the decision cannot change. Suffice it to say that the decision will inevitably lead to more detail being included in the appendix part 4 since that is the only way that main contractors can seek to impose when and where sub-contractors will work on the contract.

11.2   Extension of sub-contract time – written notice of delay

11.2   .1

The sub-contractor is under an obligation to give written notice if and whenever the commencement, progress or completion of the sub-contract works or any part thereof is affected as stated. The sub-contract works can be affected by any reason and irrespective of that reason, even if it is due to default or effect of the sub-contractor himself, the sub-contractor is obliged to write to the contractor with full details of the *material circumstances giving rise to the notice including, in so far as the sub-contractor is able, the cause or causes of the delay and identifying any matter which in his opinion comes within clause 11.3.1 (act, omission or default of the contractor etc. or the occurrence of a relevant event).*

In the author's experience no sub-contractor complies fully with the requirements of clause 11.2.1; notices, if given at all, are invariably late and fail to give details of the material circumstances as required by the clause. Few sub-contractors, if any, are open about their own shortcomings and they will not give written notice where the cause of delay is of their own making. Rather, they will try to disguise the real cause of delay and either not give notice at all, hoping that something will happen later which will 'give them a way out', or give notice in vague terms which are not based on any fact, as a means of buying time. All this will do is lead to open warfare between contractor and sub-contractor.

It is far better to adopt an open and honest stance in respect of all delays irrespective of the cause, and to open a dialogue such that no shocks occur. All the contractor wants is a true and honest appraisal of the situation so that he can, where relevant, give due notice under the main contract or take steps to redress the situation with obvious advantages to all involved or possibly help the sub-contractor to overcome his own deficiencies. Failure to give due and proper notice will inevitably lead to dispute and possibly arbitration, which can be very expensive to fund irrespective of the final outcome.

11.2 Particulars, estimates and further written notices

11.2 .2 .1, .2, .3

The words 'the Sub-Contractor shall' occur twice in clause 11.2.2 which places an obligation on the sub-contractor to give *in writing* the details required by sub-clauses .1, .2 and .3. Note that the sub-clauses are all linked with the word 'and' and the sub-contractor must *comply with the requirements of all three sub-clauses.*

Again in the author's experience no sub-contractor complies fully with the requirements of clause 11.2.2. Notices, if given at all, are invariably late and fail to give full details of the expected effects in an attempt to keep the notice vague and open in case, presumably, something is missed. The notice thereby becomes useless as a working document, but sub-contractors should not be afraid of giving detail; clause 11.2.2.3 allows for and indeed requires updating of any material changes.

11.3 Act, omission or default of contractor or relevant event

11.3 .1, .2

The sub-contractor has the right to due and proper consideration and award of an extension of time by the contractor, but only if the sub-contractor has given due notice under clause 11.2 and not further.

Note that any award of an extension of time is a decision for the contractor only and is not dependent upon the contractor receiving an award under the main contract, but any extension of time to be given is only what the contractor then estimates to be reasonable. One cannot overstress the importance of adequate and detailed notice complying with the provisions of clause 11.2; in order for the sub-contractor to exercise his right to an award of an extension of time, the sub-contractor must have fulfilled his obligations as detailed in clause 11.2. Equally the more detailed the information given, the easier it will be for the contractor to agree to the presented case, award the extension and prevent a dispute occurring, but *if notice is not given the contractor is under no obligation to award an extension of time under the provisions of this clause.*

11.4 Time limit for fixing the revised period or periods for the sub-contract works

11.4 .1 .1, .2

The sub-contractor has the right to a revised period or periods within the time scale specified but only if reasonably practicable *having regard to the sufficiency of the notice, particulars and estimates.*

Comment has been made under clause 11.3 above as to the importance of adequate and detailed notice and this clause reinforces that comment; vague, unsubstantiated statements purporting to give notice may well be rejected as being insufficient as regards notice and/or particulars and/or estimates and, if correct, the sub-contractor would lose his right to an award within the time limits specified.

11.4 .2 .1, .2

In the event of an award being made the sub-contractor has the right to know which of the various events notified have been taken into account in arriving at the award.

11.5 Extension of sub-contract time

The sub-contractor has the right to be told, in writing and not later than the period or periods specified in the clause, if the contractor is unable to give an extension of time.

11.6 Omission of work

The contractor is obliged to take account of variations omitting work etc. in arriving at an award of an extension of time, but only after a previous revision to the period or periods for completion of the sub-contract works and only if the variation was issued after the last occasion when a revision to the period or periods was made, i.e. a variation omitting work prior to a delay occurring cannot subsequently be used to offset the effects of the later delay.

11.7   Review of period for completion of sub-contract works

Note the use of the word 'may'; the contractor is not obliged to carry out what follows in this clause and the sub-contractor has, therefore, no right to a review before the date of practical completion of the sub-contract works established under clause 14.1 or clause 14.2.

The later use of the word 'shall', however, places an obligation on the contractor, and the sub-contractor has a right to a review of the period for completion of the sub-contract works no later than 16 weeks from the expiry of the practical completion of the sub-contract works, but only in so far that practical completion has been established under clause 14.1 or clause 14.2.

11.7   .1
Attention was drawn to the importance of notice in the commentary under clause 11.3 above and it was noted that under clause 11.3 the sub-contractor has the right to due and proper consideration and award of an extension of time by the contractor *but only to the extent that the sub-contractor has given due and proper notice under clause 11.2* and not further. Clause 11.7 obliges the contractor, and the sub-contractor therefore has a right, to due and proper consideration of items *not specifically notified by the sub-contractor under clause 11.2* as well as a review of items properly notified.

11.7   .2
Note that the contractor can only fix a shorter period than that previously fixed in the event that a variation omitting work has been issued subsequent to the last occasion when the sub-contract period was revised. He cannot as a part of his general review fix a period which is shorter than that previously fixed.

11.7   .3
If no alteration to the period or periods for completion of the sub-contract works is to be made, the contractor must

confirm to the sub-contractor the period or periods previously fixed and the sub-contractor has the right to such confirmation.

11.8 Sub-contractor's best endeavours to prevent delay

This clause contains the word 'shall' three times which places an obligation on the sub-contractor to do that which is listed in the clause. The opening words are also of extreme importance – 'The operation of clause 11 shall be subject to . . .' – and mean that everything within clause 11 is overridden by this clause which obliges the sub-contractor to prevent delay in the progress of the sub-contract works and to do all that may reasonably be required to proceed with the sub-contract works.

The decision in the case of *Piggott Foundations Limited* v. *Shepherd Construction Limited* (1993) to a large extent was determined by the opening words of this clause, namely that the operation of clause 11 was subject to the sub-contractor preventing delay in the progress of, and doing all that was reasonably required to proceed with, the sub-contract works. As such the imposition of constraints beyond that laid down in the appendix regarding order of work etc. was held to be contractually invalid and the sub-contractor, therefore, had the right to carry out the sub-contract works in whatever order he wished.

11.9 Limitation on power to fix shorter period or periods for completion of the sub-contract works

The sub-contractor has the right, as a minimum, to the period or periods included in the appendix part 4 and no reduction to the period or periods stated therein can be made even if the extent of the sub-contract works is reduced by way of variations omitting work after the sub-contract has been executed.

11.10 Relevant events

.1 to .14 inclusive
These sub-clauses mirror the relevant events listed in the main contract clause 25.4, but with clause 25.4.13 numbered as 11.10.14 and a totally unique (to DOM/1) sub-clause

11.10.13. Clause 11.10 does not give any rights to or place any obligations on the sub-contractor; it is merely a list of events which may give rise to an extension of time under clause 11 but only provided that all of the requirements of clause 11 have been complied with by the sub-contractor.

12 Failure of sub-contractor to complete on time

12.1

If the sub-contractor fails to complete, the contractor is obliged to notify the sub-contractor, and the sub-contractor, therefore, has a right to receive the said notice.

12.2

The sub-contractor is obliged to pay or allow any sums due under this clause but only on receipt of the notice in clause 12.1.

The issue and receipt of this notice is, therefore, essential if the contractor intends to pursue direct loss and expense suffered or incurred by the contractor as provided in clause 12.2. Without the notice the sub-contractor is not under any obligation to pay or allow any loss and/or expense.

13 Matters affecting regular progress – direct loss and/or expense – contractor's and sub-contractor's rights

13.1 Disturbance of regular progress of sub-contract works – sub-contractor's claims

The sub-contractor has a right to recover the *agreed* amount of any direct loss and/or expense caused to the sub-contractor for the reasons as listed in the clause, *and for those reasons only*, but only if the sub-contractor makes written application in accordance with the requirements of the clause and thus the sub-contractor is obliged to make such written application.

The grounds for recovery of loss and expense are severely restricted to deferment (and then only if clause 23.1.2 of the main contract conditions applies), default of the contractor etc. or the relevant matters listed and sub-contractors must be aware that any claim under the second head (default of the contractor etc.) will be resisted, if not by the contractor then certainly by any other sub-contractor involved. Claims against the relevant matters will, because they can in effect be 'passed on', at least be heard but, whatever the grounds, sub-contractors should not delay in giving written notice for 'commercial considerations': the clause obliges the sub-contractor to make written application 'within a reasonable time of such material effect becoming apparent', and late submission of notices will cause more problems in having them properly considered apart from being a direct breach of the requirements of this clause.

13.1   .1, .2, .3

The three provisos to clause 13.1 are all linked by the word 'and'. They all place obligations on the sub-contractor to act as stated therein and it is worth discussing those obligations in a little more detail.

Clause 13.1.1 obliges the sub-contractor to make written application as soon as it has become, *or should reasonably have become*, apparent that the regular progress of the sub-contract works has been or is likely to be affected and clearly this reinforces the words in the opening paragraph to clause 13.1 to make written application 'within a reasonable time of such material effect becoming apparent'.

Note the words in bold in the last sentence – 'or should reasonably have become'; the whole purpose of clause 13 is to make sub-contractors and main contractors communicate to prevent late notice of matters causing loss and/or expense. The whole question of such loss and/or expense is, thereby, brought into the open whilst it is fresh in everybody's mind, such that the various arguments can be put forward and resolved based around the facts pertinent at the time when the issues were fresh and not much later when those issues have been, at best, greyed by the passage of time. The words are, therefore, sensible and should make the non-giving of notice for commercial or other considerations a 'non-starter'.

Clause 13.1.2 further obliges the sub-contractor to submit such information in support of his application as is reasonably necessary to show that regular progress has been or is likely to be affected.

As stated above, clauses 13.1.1 and 13.1.2 are linked by the word 'and' and the sub-contractor has, therefore, to comply with both sub-clauses without further reference or requests from the contractor.

Much has been written in technical papers of 'cause and effect' and 'global claims' following court decisions in *J Crosby & Sons Limited* v. *Portland Urban District Council* (1966), *London Borough of Merton* v. *Leach* (1985), *Mid Glamorgan County Council* v. *J Devonald Williams & Partners* (1992) and *McAlpine Humberoak Ltd* v. *McDermott International Inc.* (1992). Over the years these cases have shown a gradual shift away from the global or 'rolled up' claim towards more detailed particulars to show cause and effect.

The question of the extent to which a contractor may request particulars to show cause and effect is one which will no doubt continue to be debated, but in the event of a dispute arising from a claim for loss and expense under a contract, the claimant will need to prove facts which will include:

(1) the existence of the contract
(2) the occurrence of an event giving rise to an entitlement to recover direct loss and/or expense under the contract
(3) that the event has caused that loss and/or expense
(4) the extent of the loss suffered as a result of the event
(5) that the loss arises naturally and in the normal course of things from the event (direct loss and/or expense).

This is precisely what is required under this clause and sub-contractors must be prepared to provide the details. It is not sufficient to make reference to a programme; a programme is not generally a contract document – it is merely a statement of intent. To comply with the requirements of this clause it will be necessary to show clearly what work was in progress when the particular matter manifested itself and what the effect was on that work in progress, which may include, but not be limited to, total stoppage, partial stoppage, redeployment, reduced output, new or amended resources or any combination of any of these.

Clause 13.1.3 obliges the sub-contractor to submit to the contractor such details of such loss and/or expense *as the Contractor requests* in order reasonably to enable that direct loss and/or expense to be agreed.

Again this is linked by the word 'and', but unlike clauses 13.1.1 and 2, this clause is subject to a contractor's request. The sub-contractor is not under any obligation to provide details initially, although as much detail as possible initially will prove helpful in agreeing the amount of the loss and/or expense, but when requested by the contractor, the sub-contractor must comply with that request. It is submitted that a request to 'provide details of your loss and/or expense' would be insufficient to comply with the requirements of this clause; the intention is for the contractor to make requests 'in order reasonably to enable that direct loss and/or expense to be agreed' and, as such, those requests must be specific. Anything other than a specific request for information/details should be regarded as a delaying tactic, although one should also have regard to the fact that, if it is for one of the relevant matters referred to in clause 13.3, the request is likely to have been orchestrated by the quantity surveyor named in the main contract, but again his requests for information should be specific and not general.

The sub-contractor does not have any right to a request for details of loss and/or expense; any request is purely to enable agreement of the loss and expense, although failure by the contractor to request details could be interpreted as acceptance of the details put forward by the sub-contractor initially, notwithstanding that no time scale is stated as to when any request should be made.

13.2 Relevance of certain extensions of sub-contract time

The contractor is obliged to state in writing and the sub-contractor has, therefore, a right to written notice specifying what extension of time has been made under the clauses as listed, *but purely to the extent that it is necessary for the agreement of any loss and/or expense applied for under clause 13.1.*

Reference and emphasis has been made previously to the fairly onerous obligations of the sub-contractor under clause 13.1 and that failure to comply with those terms could place the recovery of loss and/or expense at risk. Equally here, failure to comply fully with the requirements of clause 13.1 could remove the sub-contractor's right under this clause.

13.3 Relevant matters

.1 to .7 inclusive
These sub-clauses merely list those matters which can give rise to the recovery of direct loss and/or expense referred to in clause 13.1 and contain no additional rights or obligations over and above those listed in clauses 13.1 and 13.2.

Sub-contractors should never attempt to claim for items which do not fall under one of the specified 'heads of claim' as a means to recover bad estimating or other loss reasons. Such a course of action would undoubtedly fail, although the author has come across 'claims' disguised to coincide with one of the relevant matters; all they involve is a lot of time, effort and cost.

13.4 Disturbance of regular progress of works – contractor's claims

The contractor has a right to recover the *agreed* amount of any direct loss and/or expense caused by the sub-contractor for the reasons as listed, *and for those reasons only*, but only if the contractor makes written application in accordance with the requirements of the clause and thus the sub-contractor has a right to receive such written application.

The grounds for recovery of loss and expense by the contractor are severely restricted to default of the sub-contractor, his servants or agents, and the clause obliges the contractor to make written application 'within a reasonable time of such material effect becoming apparent': late submission of notices by the contractor should be strongly contested by sub-contractors as being a direct breach of the requirements of this clause.

13.4   .1, .2, .3

The three provisos to clause 13.4 are all linked by the word 'and'. They all place obligations on the contractor to act as stated therein and the sub-contractor, therefore, has a right to those actions. The situation here is the exact reverse of the obligations of the sub-contractor in clause 13.1 and it is worth discussing these rights and obligations in a little more detail.

Clause 13.4.1 obliges the contractor to make written application as soon as it has become, *or should reasonably have become*, apparent that the regular progress of the works has been or is likely to be affected and clearly this reinforces the words in the opening paragraph to clause 13.4 to make written application 'within a reasonable time of such material effect becoming apparent'.

Note again the words emphasised in the last sentence – 'or should reasonably have become'; the whole purpose of clause 13 is to make sub-contractors and main contractors communicate to prevent late notice of matters causing loss and/or expense. The whole question of such loss and/or expense is, thereby, brought into the open whilst it is fresh in everybody's mind such that the various arguments can be put forward and resolved based around the facts pertinent at the time when the issues were fresh and not much later when those issues have been, at best, dimmed by the passage of time.

Clause 13.4.2 further obliges the contractor to submit such information in support of his application as is reasonably necessary to show that regular progress has been or is likely to be affected.

As stated above, clauses 13.4.1 and 13.4.2 are linked by the word 'and' and the contractor has, therefore, to comply with both sub-clauses without further reference or requests from the sub-contractor. Reference has been made under clause 13.1 to 'cause and effect' but that is precisely what is required under this clause. Again, it is not sufficient to make reference to a programme, and to comply with the requirements of this clause; it will be necessary to show clearly what work was in progress when the particular matter manifested itself and what the effect was on that work in progress, which may include, but not be limited to, total stoppage, partial stoppage, redeployment, reduced output, new or amended resources or any combination of any of these.

Clause 13.4.3 obliges the contractor to submit to the sub-contractor such details of such loss and/or expense *as the sub-contractor requests* in order reasonably to enable that direct loss and/or expense to be ascertained and agreed.

Again this is linked by the word 'and', but unlike clauses 13.4.1 and 2, this clause is subject to a sub-contractor's request. The contractor is not under any obligation to provide details, although as much detail as possible initially will prove helpful in agreeing the amount of the loss and/or expense, but when requested by the sub-contractor, the contractor must comply with that request. It is submitted that a request to 'provide details of your loss and/or expense' would be insufficient to comply with the requirements of this clause; the intention is for the sub-contractor to make requests 'in order reasonably to enable that direct loss and/or expense to be ascertained and agreed' and, as such, those requests must be specific.

Anything other than a specific request for information/details could be regarded as a delaying tactic, but the sub-contractor is not under any obligation to make any request; any request is purely to enable ascertainment and agreement, although failure to request details could be interpreted as acceptance of the details put forward by the contractor notwithstanding that no time scale is stated as to when any request should be made.

One cannot over-emphasise the intention behind clause 13, to bring to the attention of both parties to the sub-contract any matters which have caused direct loss and/or expense to one or the other such that that loss and/or expense can be ascertained and agreed whilst the events are fresh and the facts can be established. Following as closely as possible the rules laid down in clause 13 will enable that ascertainment and agreement to be made without the acrimony and expense which is currently prevalent in disputes within the construction industry.

13.5 Preservation of rights and remedies of contractor and sub-contractor

This is the sub-contract equivalent of clause 26.6 of the main contract and allows the sub-contractor to retain a right to make a claim for damages as an alternative to and independent of a claim for direct loss and/or expense.

This was considered in the case concerning *London Borough of Merton* v. *Stanley Hugh Leach Limited* (1985), which involved a contract carried out under the provisions of the standard form of building contract 1963, but which is still considered applicable case law to JCT 80 and DOM/1. Following an interim award from the arbitrator, Merton appealed under the Arbitration Act 1979 and the court was asked to rule on certain points of law, one of which was:

If the contractor proves all the breaches alleged (as pleaded) will he be entitled to recover sums otherwise than in accordance with clauses 11(4) and (6), 24(1) and 30 of the contract:
(a) in respect of the contract works as a whole
(b) in respect of all breaches
(c) in respect of any individual breach and if so what?

(Under the 1963 Standard Form of Contract, clause 11 relates to variations, provisional and prime cost sums, clause 24 to loss and expense caused by disturbance of regular progress of the works and clause 30 to certificates and payments. The equivalent clauses in JCT 80 are 13, 26 and 30.)

It was held that the contract was not exhaustive of the contractor's remedies. Clause 24(2) (JCT 80 equivalent clause 26.6) entitled the contractor to make a claim for damages as an alternative to or independently of a claim under clause 24(1) (JCT 80 equivalent clause 26.1) and similarly clause 11(6) (JCT 80 equivalent clause 26.2.7) does not exclude a claim for damages. Thus the contractor was not obliged to make application under clause 24(1) (JCT 80 equivalent clause 26.1) but can prefer to wait until completion of the work and join the claim for damages with other claims for damages.

Thus, following the contractual chain, the sub-contractor has an equivalent right provided by the wording of clause 13.5.

14 Practical completion of sub-contract works – liability for defects

14.1 Date of practical completion of sub-contract works

The sub-contractor is obliged to notify the contractor in writing as stated and, in the event of dissent to his notice, has the right to be notified of the reasons for such dissent.

In all his years within the industry the author has only very rarely come across sub-contractors who comply with this clause. Sub-contractors should realise that, once notice is given and practical completion of the sub-contract works is established, this gives the sub-contractor the right to release one half of retention monies in accordance with clause 21.5.2; it does not however trigger the date for commencement of the defects liability period in respect of the sub-contract works; that commences on practical completion of the works in accordance with clause 17 of the main contract.

14.2 Deemed practical completion of sub-contract works

The provisions of clause 14.2 only apply in the event that the contractor gives written notice of dissent under clause 14.1. Reference is drawn to footnote [i] which confirms that whilst there is no requirement under clause 14.2 for any agreement to be in writing, the benefit of the joint names policy of insurance granted under clause 8 ceases upon the earlier of the 'Terminal dates' as defined in clause 6.1.2, one of which is 'the date upon which the Contractor issues in writing to the Sub-Contractor a confirmation of the agreement under clause 14.2'.

For the purposes of the insurance provisions, therefore, the sub-contractor has the right to written confirmation of any agreement reached, notwithstanding the provisions of clause 14.2.

It should also be noted that the sub-contract does not envisage that sub-contractors will not comply with the requirements of clause 14.1; there is no statement regarding deemed practical completion in the event that the notice required by clause 14.1 is not given although clearly once the works are complete the sub-contract works must equally be complete.

In the recent case of *Vascroft (Contractors) Ltd* v. *Seeboard plc* (1996), which concerned a sub-contract under DOM/2 but which is equally applicable to DOM/1, following an appeal from an earlier arbitration, the courts were asked to rule on a point of law:

If the sub-contractor gives no notification in accordance with clause 14.1 of the standard form of Sub-Contract DOM/2, is practical completion deemed to have occurred in the manner set out in clause 14.2 of DOM/2? If not, is practical completion a matter of fact?

The answer to the first question was No and that to the supplementary question Yes. It was held that:

(1) Clause 14.1 imposed an obligation on the sub-contractor to 'notify the Contractor in writing when in his opinion the Sub-Contract Works are practically completed'.
(2) If the sub-contractor fails to comply with his obligation, the machinery in clause 14 cannot come into operation and no contractual means exist to establish a firm date for practical completion other than by arbitration.
(3) If the sub-contractor is in breach of his obligation under clause 14.1, the date for practical completion falls to be established as a matter of fact.

Thus it can be seen that the obligation imposed by clause 14.1 should not be taken lightly and sub-contractors should ensure that they comply with the requirements contained therein, since, in the event of a dispute as to practical completion, arbitration is the only method of establishing a firm date.

14.3 Liability of sub-contractor for defects in sub-contract works

This clause obliges the sub-contractor to take responsibility for any defects etc. which occur due to faulty workmanship and/or materials or due to frost occurring before practical completion of the sub-contract works.

There are some points regarding completion which are worthy of expansion:

(1) Clause 18 of the main contract covers partial possession by the employer and clearly the sub-contractor should be aware of any such possession since it affects his insurance requirements under the sub-contract and triggers the start of the defects liability period specified in the main contract for that part.

(2) The 'without prejudice' statement prior to the liability statement in this clause is to ensure that the sub-contractor's liability for defects is no greater than that of the contractor under the main contract. Without this statement the sub-contractor would have a never-ending liability for defects, whereas with it his liability is limited to the defects liability period specified in the main contract.

(3) The sub-contractor is only liable for defects, shrinkages and other faults which occur due to *materials or workmanship* not in accordance with the sub-contract. The author has come across many instances of alleged defects which have arisen either as a result of design or were inherent within the material specified and, under these or similar circumstances, the sub-contractor has no liability to make good these defects at his own cost. Should this situation arise the sub-contractor should record with the contractor in writing the reasons why he does not believe that the defect, shrinkage or other fault is his responsibility.

(4) Defects due to frost occurring before practical completion are, in the author's experience, very rare, but note the words in the clause – frost occurring before the date of practical completion of the *sub-contract works*. It is possible for frost to affect the sub-contract

works after they are practically complete and it is, therefore, essential, in order to protect the sub-contractor from any future liability, to comply with the requirements of clause 14.1 and notify the contractor of the date when, in the sub-contractor's opinion, the sub-contract works are practically completed.

14.4 Architect's instructions – clauses 17.2 and 17.3 of the main contract conditions

Clauses 17.2 and 17.3 of the main contract conditions give the architect the right to make an appropriate deduction in respect of any defects, shrinkages or other faults not made good and clause 14.4 of the sub-contract steps down that right and obliges the sub-contractor to accept his share of the deduction.

Such a course of action is extremely rare as architects generally require defects etc. to be made good, but there can be instances where access for the making good can be virtually impossible or the act of making good would disrupt operations to such an extent that the costs to the employer would be prohibitive. Under such circumstances an appropriate deduction would be the only sensible solution and hence this clause.

Clauses 17.2 and 17.3 of the main contract conditions do not appear to limit the architect's rights in any way, but do not, in the author's view, give the architect the right to make a deduction for all defects – such an action would be unreasonable; nor do they specify what an 'appropriate deduction' is or how this is to be calculated. This can only be assessed based on the likely costs of making good and would be subject to agreement by all parties to the contract or sub-contract as appropriate.

15 Price for sub-contract works

15.1 Sub-contract sum

The sub-contractor has the right to the sub-contract sum or such other sum as specified upon completion of the sub-contract works as the price for the sub-contract works which would subsequently become payable in accordance with the payment provisions of the sub-contract (see later clause 21).

If article 2.1 applies, the sub-contract is a lump-sum contract only to be adjusted for variations in accordance with the definitions and rules contained within the sub-contract. It follows, therefore, that nobody has the absolute right to remeasure the sub-contract works, but if such a course of action is considered appropriate as being the only realistic way of deriving the quantity of work carried out, one should then consider whether the scope of the sub-contract works has changed so much as to make the rates and prices in the sub-contract inappropriate for those works actually carried out. (See also comments under article 2 on page 14.)

15.2 Tender sum – ascertained final sub-contract sum

The sub-contractor has the right to complete re-measurement of the sub-contract works and valuation in accordance with the rules in the sub-contract as the price for the sub-contract works which would subsequently become payable in accordance with the payment provisions of the sub-contract (see later clause 21).

16 Valuation of variations and provisional sum work

16.1 Valuation

The sub-contractor has the right to have all variations and all work executed as described valued, but only where clause 15.1 applies. That valuation shall, unless otherwise agreed by both parties, be made in accordance with the provisions of clause 16 and the sub-contractor has both a right to and an obligation to abide by those rules.

The sub-contractor does not have a right to depart from the rules; only in the event of agreement by *both* parties can departure from the rules be envisaged.

16.2   Sub-contractor's schedule of rates or prices

The sub-contractor has both a right to use of and an obligation to use any rates or prices as described to the extent that those rates or prices are included in the sub-contract documents.

The sub-contract documents are defined in clause 1.3 as the Sub-Contract DOM/1 and the numbered documents and any rates and prices which are not incorporated therein cannot, as of right or obligation, be used for valuations under clause 16.3.

16.3   Valuation rules

16.3   .1

The sub-contractor has both a right and an obligation to have work measured and valued in accordance with the rules, but only to the extent that the valuation relates to additional or substituted work *which can be properly valued by measurement.*

Measurement in this context, as confirmed in the later clause 16.3.3.1, relates to measurement following the same principles as those used in the preparation of the bills of quantities. If the particular additional or substituted work cannot be measured in accordance with those principles, then the provisions of clause 16.3.1 cannot apply and the sub-contractor must look to the other rules to determine which one should apply.

There follow sub-clauses .1, .2 and .3 which set out the rules to be adopted for valuation under the particular conditions appertaining, all of which contain both a right and an obligation to abide by those rules.

16.3   .1   .1

Sub-contractors should note that all elements must be satisfied for the rule to apply. The work must be of similar character *and* executed under similar conditions *and* not significantly changed in quantity. If any one element does not apply then this rule does not apply.

16.3   .1   .2

This is the next logical step after sub-clause .1, but again all required elements must be satisfied. It must be of similar character but *not* executed under similar conditions *and/or* significantly changed in quantity. Two elements only need be satisfied here; if they are not, then this rule cannot be used.

16.3   .1   .3

Again the next logical step after sub-clause 2; the whole of clause 16.3 is logical in the way that it is set out such that once all parameters are satisfied, that is the rule to apply. In this case only one element has to be satisfied, i.e. that the work is not of similar character.

Provided the opening to clause 16.3.1 is satisfied, i.e. that the work can be properly valued by measurement, one of the sub-clauses .1, .2 or .3 must contain the rule that is to apply to the valuation of the variation.

16.3   .2

The sub-contractor has both a right and an obligation to have omitted work measured and valued in accordance with this rule, but only to the extent that the valuation relates to *the omission of work set out in bills of quantities and/or other documents.*

Varied work which is subsequently omitted is not governed by this rule. The author is aware of circumstances where, subsequent to preparation of bills of quantities, drawings have been amended prior to work starting on site and then work omitted. The quantity surveyor has sought to measure and value work omitted from the latest drawings in accordance with this rule; clearly this is incorrect and only that measured in the bills of quantities can be omitted.

16.3   .3

Clauses 16.3.1 and 16.3.2 give a right to and place an obligation on sub-contractors to use and abide by those rules; the three sub-clauses to clause 16.3.3, which apply to all valuations under clauses 16.3.1 and 16.3.2, are no different. Both parties to the sub-contract must abide by these rules.

16.3   .3   .1

Reference was made to this clause in 16.3.1 above.

Unless the work can be properly valued by measurement, the provisions of clause 16.3.1 cannot apply and this clause sets out that those measurements shall be in accordance with the same principles as those governing the preparation of the bills of quantities comprised in the sub-contract documents. Neither party to the sub-contract can depart from those measurement rules.

16.3 .3 .2

It should be noted that percentage additions or lump sum adjustments apply to all valuations under clause 16.3.1 and 16.3.2, *including work valued at fair rates and prices under clause 16.3.1.3*, and are not restricted to those items valued in accordance with clauses 16.3.1.1 and 16.3.1.2, i.e. at rates set out in the bills of quantities or pro rata thereto.

16.3 .3 .3

This is often a difficult rule to apply since generally sub-contractors do not, and indeed are often not given the opportunity to, price preliminaries separately. Compliance with clauses 16.3.1.1 and 16.3.1.2 should not cause a problem since the bill rates will be inflated by the amount of preliminaries built into the original bid and, whilst this may not be totally satisfactory, it does give the sub-contractor a general recovery of preliminaries across all variations which should be sufficient to cover any extra costs.

Valuation under clause 16.3.1.3 can prove difficult where there is no breakdown of the preliminaries in the original bid, but, irrespective of a provided breakdown or otherwise, sub-contractors have the right to have preliminaries considered in any valuation of a variation.

16.3 .4

Note the wording; this clause relates *only* to work which *cannot be valued by measurement*. It must be repeated that there is a logical progression to the way in which the rules in clause 16.3 are set out, but often sub-contractors will turn to this clause immediately as a way of ensuring that all costs are covered. *This is wrong*; sub-contractors and contractors must follow the logical progression within clause 16.3 and use the rule which first applies to the circumstances surrounding the variation. The basic question is, 'Is it measurable?'. If the answer is yes, clause 16.3.1 or clause 16.3.2 will apply; if the answer is no, clause 16.3.4 will apply.

16.3 .4 .1, .2

There is a tendency for 'specialist trades' to believe that they fall within the province of clause 16.3.4.2, no matter what their trade is. Reference to the wording of clause 16.3.4.2 and footnote [j] clearly shows that this is not true; the only trades falling within the rule in clause 16.3.4.2 are those in which the Royal Institution of Chartered Surveyors and the appropriate body representing the employers in that trade have agreed and issued a definition of prime cost of daywork. Footnote [j] confirms those bodies, and thereby the trades they represent, as being the Electrical Contractors Association, the Electrical Contractors Association of Scotland and the Heating and Ventilating Contractors Association. All other 'specialist trades' are bound by the rules in clause 16.3.4.1.

The sub-contractor has an obligation to comply with *all* of the requirements of the proviso if he considers that valuation of work executed should be in accordance with the rules in clause 16.3.4.

All too often sub-contractors fail either in timeous presentation of records for verification or in presentation of incomplete records, which they subsequently try and recover in final account negotiations. The sub-contract unfortunately makes no provision for action in the event of failure to abide by the requirements of the proviso, nor is there any established precedent, but it is suggested that, unless the sub-contractor complies with the requirements in full, the contractor has no liability to value, and subsequently make payment, in accordance with clause 16.3.4.

16.4 .1

The whole of clause 16.4 on the face of it is very straightforward but, in the author's experience, it is very rarely used by the quantity surveyor and is subject to an overall proviso (see later).

The intention is to reflect within the value of a variation the effect of that variation on other work in so far as it substantially changes the *conditions* under which that other work, which is not the subject of the variation, is executed, and the sub-contractor has the right to have that other work treated as if it had been varied by an instruction and valued in accordance with the provisions of clause 16.

16.4   .2
This is the final rule and gives the sub-contractor the right to a fair valuation in the event that the valuation does not relate to or cannot reasonably be effected by the circumstances listed in the clause.

The overall proviso after clause 16.4.2 is of great importance and the sub-contractor is obliged to abide by the terms stated therein, i.e. that no allowance is to be made in a clause 16.3 or 16.4 valuation for delay, disruption or other disturbance costs which are reimbursable under any other provision in the sub-contract, i.e. generally clause 13 which is the provision under which the sub-contractor's claims for direct loss and/or expense would be reimbursed.

16.5   Addition to or deduction from sub-contract sum

Clause 16.5 should be read in conjunction with clause 3 and provides that any valuation under clause 16 is to be included in interim certificates and the sub-contractor has the right to be paid therefor.

17   Valuation of all work comprising the sub-contract works

17.1   Valuation of all sub-contract works

The sub-contractor has the right to have all work executed as described valued, but only where clause 15.2 applies. That valuation shall, unless otherwise agreed by both parties, be made in accordance with the provisions of clause

17.3 (and clause 37.2.3 where applicable) and the sub-contractor has both a right to and an obligation to abide by those rules.

The sub-contractor does not have a right to depart from the rules; only in the event of agreement by *both* parties can departure from the rules be envisaged.

17.2 Sub-contractor's schedule of rates or prices

The sub-contractor has both a right to use of and an obligation to use any rates or prices as described to the extent that those rates or prices are included in the sub-contract documents.

The sub-contract documents are defined in clause 1.3 as the Sub-Contract DOM/1 and the numbered documents and any rates and prices which are not incorporated therein cannot, as of right or obligation, be used for valuations under clause 17.3.

17.3 Valuation rules

17.3 .1
The sub-contractor has both a right and an obligation to have work measured and valued in accordance with the rules, but only to the extent that the valuation relates to work *which can be properly valued by measurement.*

Measurement in this context, as confirmed in the later clause 17.3.2.1, relates to measurement following the same principles as those used in the preparation of the bills of quantities: if the particular work cannot be measured in accordance with those principles, then the provisions of clause 17.3.1 cannot apply and the sub-contractor must look to the other rules to determine which one should apply.

There follow sub-clauses .1, .2 and .3 which set out the rules to be adopted for valuation under the particular conditions appertaining, all of which contain both a right and an obligation to abide by those rules.

17.3   .1   .1

Sub-contractors should note that all elements must be satisfied for the rule to apply. The work must be of similar character *and* executed under similar conditions *and* not significantly change the quantity. If any one element does not apply then this rule does not apply.

17.3   .1   .2

This is the next logical step after sub-clause .1, but again all required elements must be satisfied. It must be of similar character but *not* executed under similar conditions *and/or* it significantly changes the quantity. Two elements only need be satisfied here; if they are not then this rule cannot be used.

17.3   .1   .3

Again the next logical step after sub-clause 2. The whole of clause 17.3 is logical in the way that it is set out such that once all parameters are satisfied, that is the rule to apply. In this case only one element has to be satisfied, i.e. that the work is not of similar character.

Provided the opening to clause 17.3.1 is satisfied, i.e. that the work can be properly valued by measurement, one of the sub-clauses .1, .2 or .3 must contain the rule that is to apply to the valuation of the variation.

17.3   .2

Clause 17.3.1 gives a right to and places an obligation on sub-contractors to use and abide by those rules; the three sub-clauses to clause 17.3.2, which apply to all valuations under clauses 17.3.1, are no different. Both parties to the sub-contract must abide by these rules.

17.3   .2   .1

Reference was made to this clause in 17.3.1 above. Unless the work can be properly valued by measurement, the provisions of clause 17.3.1 cannot apply and this clause sets out that those measurements shall be in accordance with the same principles as those governing the preparation of the bills of quantities comprised in the sub-contract documents. Neither party to the sub-contract can depart from those measurement rules.

17.3   .2   .2

It should be noted that percentage additions or lump sum adjustments apply to all valuations under clause 17.3.1, *including work valued at fair rates and prices under clause 17.3.1.3*, and are not restricted to those items valued in accordance with clauses 17.3.1.1 and 17.3.1.2, i.e. at rates set out in the bills of quantities or pro rata thereto.

17.3   .2   .3

This is often a difficult rule to apply since generally sub-contractors do not, and indeed are often not given the opportunity to, price preliminaries separately.

Where the preliminaries are priced separately, this clause requires those preliminaries to be included in the valuation of work; it further requires that in respect of variations or in regard to the expenditure of provisional sums included in the sub-contract documents, the preliminaries are to be adjusted to take account of those matters and the sub-contractor has a right to due and proper consideration of preliminaries in respect of all work carried out.

Where preliminaries are not priced separately, compliance with clauses 17.3.1.1 and 17.3.1.2 should not cause a problem since the bill rates will be inflated by the amount of preliminaries built into the original bid and, whilst this may not be totally satisfactory, it does give the sub-contractor a general recovery of preliminaries across all works which should be sufficient to cover costs. Valuation under clause 17.3.1.3 can prove difficult where there is no breakdown of the preliminaries in the original bid, but, irrespective of a provided breakdown or otherwise, sub-contractors have the right to have preliminaries considered in any valuation of their work.

17.3   .3

Note the wording – this clause relates *only* to work which *cannot be valued by measurement*. It must be repeated that there is a logical progression to the way in which the rules in clause 17.3 are set out, but often sub-contractors will turn to this clause immediately as a way of ensuring that all costs are covered. *This is wrong*. Sub-contractors and contractors must follow the logical progression within clause 17.3 and use the rule which first applies to the circumstances surrounding the valuation. The basic question is, 'Is it measurable?'. If the answer is yes, clause 17.3.1 will apply; if the answer is no, clause 17.3.3 will apply.

17.3   .3   .1, .2

There is a tendency for 'specialist trades' to believe that they fall within the province of clause 17.3.3.2, no matter what their trade is. Reference to the wording of clause 17.3.3.2 and footnote [k] clearly shows that this is not true; the only trades falling within the rule in clause 17.3.3.2 are those in which the Royal Institution of Chartered Surveyors and the appropriate body representing the employers in that trade have agreed and issued a definition of prime cost of daywork. Footnote [k] confirms those bodies, and thereby the trades they represent, as being the Electrical Contractors Association, the Electrical Contractors Association of Scotland and the Heating and Ventilating Contractors Association. All other 'specialist trades' are bound by the rules in clause 17.3.3.1.

The sub-contractor has an obligation to comply with *all* of the requirements of the proviso if he considers that valuation of work executed should be in accordance with the rules in clause 17.3.3.

All too often sub-contractors fail either in timeous presentation of records for verification or in presentation of incomplete records, which they subsequently try and recover in final account negotiations. The sub-contract unfortunately makes no provision for action in the event of failure to abide by the requirements of the proviso, nor is there any established precedent, but it is suggested that, unless the sub-contractor complies with the requirements in full, the contractor has no liability to value, and subsequently make payment, in accordance with clause 17.3.3.

17.3 .4

As with clause 16.4, in the author's experience clause 17.3.4 is very rarely used by the quantity surveyor and is subject to an overall proviso (see later).

Again the intention is to reflect within the value of work executed the effect of directions on other work in so far as they substantially change the *conditions* under which that other work, which is not the subject of a direction, are executed, and the sub-contractor has the right to have that other work treated as if it had been varied by a direction and valued in accordance with the provisions of clause 17.3.1.2.

17.3 .5

This is the final rule and gives the sub-contractor the right to a fair valuation in the event that the valuation does not relate to or cannot reasonably be effected by the circumstances listed.

The overall proviso to clause 17.3 is of great importance and the sub-contractor is obliged to abide by the terms stated therein, i.e. that no allowance is to be made in a clause 17.3 valuation for delay, disruption or other disturbance costs which are reimbursable under any other provision in the sub-contract, i.e. clause 13 which is the provision under which the sub-contractor's claims for direct loss and/or expense would be reimbursed.

18 Bill of quantities – standard method of measurement

18.1 Preparation of bills of quantities – errors in preparation, etc.

18.1 .1

The first standard method of measurement was published in 1922 in recognition of the need for a uniform approach to producing bills of quantities. The current standard method is no different, providing a uniform basis for measuring building works and embodying the essentials of good practice to fully describe and accurately represent the quantity and quality of the works to be carried out. The sub-contract recognises the wisdom of this uniform approach and requires that bills of quantities shall be prepared in accordance with the *Standard Method of Measurement* 6th edition (for 7th edition see amendment 7), unless specifically stated otherwise in the bills.

In this context, bland statements such as 'The Sub-Contractor is deemed to abide by the measurement methods included within the Sub-Contract Documents' are in direct conflict with the requirements of clause 18.1 and cannot be enforced. Specific departures which draw the sub-contractor's attention to those departures such as 'Notwithstanding the requirements of clause ? of SMM6, this has been measured by etc.' do comply with the requirements of this clause and must be accepted.

18.1 .2
This clause gives the sub-contractor the right to have any departures, errors or omissions corrected; no written direction is necessary to have the relevant item treated as a variation, but the clause does not give the sub-contractor an automatic right to re-measurement in the event that clause 15.1 applies.

18.1 .3

19A Value added tax

Clause 19A applies where the sub-contractor issues receipts as referred to in Regulation 12(4) of the Value Added Tax (General) Regulations 1985 (authenticated receipts) and has not consented to tax documents prepared by the contractor under Regulations 12(3) and 26 of the Value Added Tax (General) Regulations 1985.

19A.1 .1, .2
Clause 19A.1.1 and .2 deals with the situation where exemption from VAT occurs after the date of the tender, i.e. where taxable goods and services supplied are subsequently exempt from the tax; the sub-contract price will then be inclusive of the sub-contractor's input tax and this clause gives the sub-contractor the right, in these unusual

circumstances, to be paid an amount equal to the loss which he would sustain by reason of the fact that the input tax had become irrecoverable.

19A.2

19A.3

19A.4

The contractor must pay and the sub-contractor has the right to receive any tax chargeable on any goods and services supplied by the sub-contractor to the contractor.

The sub-contractor's right under this clause is non-negotiable and arises from the sub-contractor's duty to pay and collect value added tax introduced by the Finance Act 1972, which is under the care and management of the Commissioners of Customs and Excise. Failure to collect the tax from the contractor will lead to the sub-contractor being liable to the commissioners for the tax not collected.

19A.5 .1
The appendix part 6 must state if clause 19A.5 applies or not and the sub-contractor is obliged to comply with that statement save as stated in this clause.

If clause 19A.5 applies, clause 19A.6.1 and .2 do not apply unless and until any notice issued under clause 19A.5.4 becomes effective or the sub-contractor fails to give the written notice required under clause 19A.5.2.

Footnote [1] confirms that clause 19A.5 can only apply where the sub-contractor is satisfied *at the date the sub-contract is entered into* that his output tax on *all* supplies to the contractor under the sub-contract will be at either a positive or zero rate of tax. In the author's experience all supplies will normally be positively rated.

The purpose of clause 19A.5 is to simplify the collection of and reduce the paperwork necessary to collect the tax by simply issuing one document prior to the first payment becoming due instead of a provisional assessment of tax due before each and every payment is due. It is, therefore, in the sub-contractor's interest to comply fully with the requirements of clause 19A.5 if it is stated to apply.

19A.5 .2

The sub-contractor is obliged to give written notice, not later than 14 days before the first payment under the sub-contract is due, of the rate of tax chargeable on the supply of goods and services. He is further obliged to give amendments to his written notice if the rate of tax is varied under statute, the amended notice being required to be given not later than 7 days after the date when the varied rate comes into effect.

Bearing in mind that, under clause 19A.5.1, one of the reasons given for 19A.6.1 and .2 applying is failure to give the written notice required under clause 19A.5.2, which leads to the loss of the benefit provided by clause 19A.5, it is important to comply not only with the giving of the notice but also with the time constraints stated. There is no reversal situation under clause 19A.6; once the benefits of clause 19A.5 are lost, they are lost forever under the particular sub-contract.

19A.5 .3

In order to comply with the requirements of clause 19A.2, an amount calculated at the rate given in the written notice provided under clause 19A.5 is to be added (by the contractor) to each payment due and the sub-contractor, therefore, has the right to receive the amount added as aforesaid.

19A.5 .4

Either party to the sub-contract has the right by giving a written notice, one to the other, to state that the provisions of clause 19A.5 do not apply, following which the provisions of clause 19A.6.1 and .2 will apply.

Clearly there is no benefit to the sub-contractor to change to clause 19A.6, since that will involve more paperwork, but it is not unknown for main contractors to request the change, or indeed insist upon clause 19A.6 from the start, as failure to provide the written notice required by clause 19A.6 (for each and every payment) could lead to the main contractor withholding payment of tax due, which he will have received from his client, and gaining cash flow advantage to the detriment of the sub-contractor.

19A.6 .1

The sub-contractor is obliged to give a written provisional assessment, *not later than 7 days before payment is due*, of the respective values of those supplies and goods which will be chargeable at zero rate of tax and any other rate or rates of tax. The sub-contractor is also obliged to specify the rate or rates of tax which are chargeable on non-zero rated work and the grounds upon which he considers such supplies are so chargeable.

It should be noted that the written assessment is required for each and every payment due under the sub-contract.

19A.6 .2

The contractor is required to remit such amounts of tax calculated in accordance with the assessment under clause 19A.6.1 to the sub-contractor within the period specified in clause 21 and the sub-contractor has the right to receive such tax as aforesaid.

The trigger for the calculation of tax is the sub-contractor's provisional assessment under clause 19A.6.1 and failure to provide the written assessment within the period specified will lead to non-payment of tax due by the contractor, apart from constituting an offence under the act for which the sub-contractor would become liable.

19A.7

The sub-contractor has an obligation to issue a receipt as referred to in the regulations (the authenticated receipt) *immediately* upon receipt of the amounts referred to in clause 21 and in either clause 19A.5.3 or clause 19A.6.2, whichever is applicable.

Failure to issue the relevant authenticated receipt by the sub-contractor may lead to withholding of future VAT payments – see clause 19A.10 below – and it is vital that sub-contractors comply with the requirements of this clause.

The sub-contractor must account for tax in proportion to the written provisional assessment even if the contractor does not pay the tax properly due; as stated above, it is vitally important for the sub-contractor to issue the written notice exactly as required by clause 19A.6.1 if he is not to be cash disadvantaged by non-payment of VAT.

19A.8

Any cash discount disallowed under clause 21.3 is exempt from the requirements of clause 19A.5.3 or clause 19A.6.2, whichever is appropriate, and no written notice is required to be issued by the sub-contractor. The contractor will not pay tax on the disallowed discount and the sub-contractor will not have to account to the commissioners for any output tax on those payments.

19A.9

The sub-contractor is obliged to notify the contractor and the contractor is obliged to make any necessary adjustment if the amount paid under clause 19A.5.3 or clause 19A.6.2 as applicable is not the amount of tax properly chargeable.

19A.10

If the contractor has not received from the sub-contractor, within the period specified, the receipt required under clause 19A.7, the contractor is entitled to withhold further payments (in respect of tax only), but only if the contractor gives written notice to the sub-contractor of his intent, and the sub-contractor has, therefore, the right to receive such notice in the circumstances stated.

The issue and receipt of the notice referred to is a prerequisite for action under this clause.

19B **Value added tax – special arrangement – Value Added Tax Act 1983 s.5(4) – VAT (General) Regulations 1985, Regulations 12(3) and 26**

Immediate attention is drawn to footnote [m] which confirms that clause 19B can *only* apply where the contractor has been allowed to prepare tax documents and the sub-contractor has consented to tax documents prepared by the contractor under Regulations 12(3) and 26 of the Value Added Tax (General) Regulations 1985.

19B.1   .1, .2

Clause 19B.1.1 and .2 deals with the situation where exemption from VAT occurs after the date of the tender, i.e. where taxable goods and services supplied are subsequently exempt from the tax. The sub-contract price will then be inclusive of the sub-contractor's input tax and this clause gives the sub-contractor the right, in these unusual circumstances, to be paid an amount equal to the loss which he would sustain by reason of the fact that the input tax had become irrecoverable.

19B.2

19B.3

19B.4

The contractor must pay, and the sub-contractor has the right to receive, any tax chargeable on any goods and services supplied by the sub-contractor to the contractor.

The sub-contractor's right under this clause is non-negotiable and arises from the sub-contractor's duty to pay and collect value added tax introduced by the Finance Act 1972, which is under the care and management of the Commissioners of Customs and Excise. Failure to collect the tax from the contractor will lead to the sub-contractor being liable to the commissioners for the tax not collected.

19B.5   .1

The appendix part 6 must state if clause 19B.5 applies or not and the sub-contractor is obliged to comply with that statement save as stated in this clause.

If clause 19B.5 applies, clauses 19B.6.1 and .2 do not apply unless and until any notice issued under clause 19B.5.4 becomes effective or the sub-contractor fails to give the written notice required under clause 19B.5.2.

Footnote [n] confirms that clause 19B.5 can only apply where the sub-contractor is satisfied *at the date the sub-contract is entered into* that his output tax on *all* supplies to the contractor under the sub-contract will be either a positive or zero rate of tax. In the author's experience all supplies will normally be positively rated.

The purpose of clause 19B.5 is to simplify the collection of and reduce the paperwork necessary to collect the tax by simply issuing one document prior to the first payment becoming due instead of a provisional assessment of tax due before each and every payment is due. It is, therefore, in the sub-contractor's interest to comply fully with the requirements of clause 19B.5 if it is stated to apply.

19B.5 .2

The sub-contractor is obliged to give written notice, not later than 14 days before the first payment under the sub-contract is due, of the rate of tax chargeable on the supply of goods and services. He is further obliged to give amendments to his written notice if the rate of tax is varied under statute, the amended notice being required to be given not later than 7 days after the date when the varied rate comes into effect.

Bearing in mind that, under clause 19B.5.1, one of the reasons given for 19B.6.1 and .2 applying is failure to give the written notice required under clause 19B.5.2, which leads to the loss of the benefit provided by clause 19B.5, it is important to comply not only with the giving of the notice but also with the time constraints stated. There is no reversal situation under clause 19B.6; once the benefits of clause 19B.5 are lost, they are lost forever under the particular sub-contract.

19B.5 .3

In order to comply with the requirements of clause 19B.2, an amount calculated at the rate given in the written notice provided under clause 19B.5 is to be added (by the contractor) to each payment due and the sub-contractor, therefore, has the right to receive the amount added as aforesaid.

19B.5 .4

Either party to the sub-contract has the right by giving a written notice, one to the other, to state that the provisions of clause 19B.5 do not apply, following which the provisions of clause 19B.6.1 and .2 will apply.

Clearly there is no benefit to the sub-contractor to change to clause 19B.6, since that will involve more paperwork, but it is not unknown for main contractors to request the change, or indeed insist upon clause 19B.6 from the start, as failure to provide the written notice required by clause 19B.6 (for each and every payment) could lead to the main contractor withholding payment of tax due, which he will have received from his client, and gaining cash flow advantage to the detriment of the sub-contractor.

19B.6 .1
The sub-contractor is obliged to give a written provisional assessment, *not later than 7 days before payment is due*, of the respective values of those supplies and goods which will be chargeable at zero rate of tax and any other rate or rates of tax. The sub-contractor is also obliged to specify the rate or rates of tax which are chargeable on non-zero rated work and the grounds upon which he considers such supplies are so chargeable.

It should be noted that the written assessment is required for each and every payment due under the sub-contract.

19B.6 .2
The contractor is required to remit such amounts of tax calculated in accordance with the assessment under clause 19B.6.1 to the sub-contractor within the period specified in clause 21 and the sub-contractor has the right to receive such tax as aforesaid.

The trigger for the calculation of tax is the sub-contractor's provisional assessment under clause 19B.6.1 and failure to provide the written assessment within the period specified will lead to non-payment of tax due by the contractor apart from constituting an offence under the act for which the sub-contractor would become liable.

19B.7 .1
The contractor is obliged to issue, and the sub-contractor, therefore, has the right to receive, together with the payment under clause 21 and the payment of tax under clause 19B.5.3 or 19B.6.2, whichever is applicable, a document approved by the commissioners under the regulations specified.

The document must contain the *date of despatch only* of the document to the sub-contractor, completed by the contractor – there must be no reference to any date or other matter which purports to represent for any purposes the time of supply in respect of which the sub-contractor would become liable for output tax – and space must be left for completion by the sub-contractor of the date of receipt of the document by the sub-contractor. The dates are important as they determine when the sub-contractor is liable to the commissioners for payment of output tax.

If the payment has not been received by the sub-contractor as stated in the clause, the sub-contractor is obliged to reject the document immediately and explain to the contractor the reasons for such rejection.

19B.7 .2
Any cash discount disallowed under clause 21.3 is exempt from the requirements of clause 19B.5.3 or clause 19B.6.2, whichever is appropriate, and no written notice is required to be issued by the sub-contractor. The contractor will not pay tax on the disallowed discount and the sub-contractor will not have to account to the commissioners for any output tax on those payments.

19B.7 .3
The contractor is obliged to issue, and the sub-contractor, therefore, has the right to receive, a reconciliation statement with the document if, and only if, the payment is different to that stated in the accompanying document.

19B.8 .1, .2
The sub-contractor is obliged to abide by the provisions of clause 19A in the event that the commissioners withdraw the approval referred to in clause 19B.7.1 or the sub-contractor withdraws his consent to the procedure referred to in clause 19B.7.1.

In this case the sub-contractor will have to issue authenticated receipts for each and every payment received under the sub-contract.

19B.9

The sub-contractor must not issue any document which is or purports to be an authenticated receipt unless the provisions of clause 19B.8, and thereby clause 19A, apply. If he does there will be two tax documents in existence and the sub-contractor may be held to be liable for both tax amounts as output tax.

19B.10

20A   Finance (No.2) Act 1975 – Tax deduction scheme

Immediate attention is drawn to footnotes [o] and [p].

Footnote [o] refers to the fourth recital (A) which confirms that if the sub-contractor is a user of a current sub-contractor's tax certificate, i.e. the words 'is not' are deleted, the provisions of clause 20A apply to the sub-contract.

Footnote [p] confirms that, where the contractor deals with the tax deduction scheme by way of an Inland Revenue approved self-vouchering system, clause 20A is not applicable and both the contractor and sub-contractor must make appropriate sub-contractual arrangements.

The sub-contractor is, therefore, obliged under these circumstances to make such arrangements; those arrangements will then take the place of the standard clauses and may or may not give the sub-contractor rights or obligations dependent upon the arrangements agreed.

20A.1   Definitions

20A.2   .1
Production of tax certificates

20A.2   .1   .1

The sub-contractor is obliged to produce to the contractor his current tax certificate not later than 21 days before the first payment is due. The contractor is obliged to confirm in writing, and the sub-contractor has, therefore, the right to receive confirmation of, the production and his satisfaction or non-satisfaction of the tax certificate under the regulations within 7 days of production as aforesaid.

20A.2   .1   .2
Certificate holder – production of document

The sub-contractor is obliged to lodge with the contractor a document (form numbered 714C) as referred to in the regulations not later than 21 days before the first payment is due. The contractor is obliged to confirm in writing, and the sub-contractor has, therefore, the right to receive confirmation of, the production and his expression of doubt or otherwise as to the correctness of the information shown on the document within 7 days of production as aforesaid.

In the author's experience contractors rarely, if ever, provide the written confirmation required by this clause. Non-payment of certificates under clause 21, for reasons such as 'non-receipt of tax documents' or 'unhappy with tax documents', are not unknown and sub-contractors should, therefore, ensure that they receive written confirmation of their tax documentation as required by this clause.

Sub-contractors also believe that when their tax certificate or document expires it is the contractor's responsibility to bring this to the sub-contractor's attention and request a replacement. It is suggested that this is incorrect; production of tax documentation is solely the responsibility of the sub-contractor and it follows, therefore, that it must be his responsibility to ensure that the information lodged with the contractor is up to date.

20A.2   .2
Run-on provisions

The operation of clause 20A.2.1 is subject to clause 20A.2.2 and, provided the contractor has previously expressed in writing to the sub-contractor his satisfaction in respect of either of the documents referred to in clause 20A.2.1, there is no need for the sub-contractor to produce the documentation again.

This is yet another reason to obtain the confirmation in writing required by clause 20A.2.1 since this gives the sub-contractor the right to rely on the previous production of documents as evidence of tax status for current or anticipated sub-contracts, provided of course that they are not date expired.

20A.2 .3

Change in nominated bank accounts

If clause 20A.2.1.2 applies and the sub-contractor changes the nominated bank account(s) specified on the document referred to therein, the sub-contractor is obliged to notify the contractor of such change.

20A.2 .4

Payment without statutory deduction

Where the contractor has given the confirmation in writing required by clause 20A.2.1.1 or 20A.2.1.2 as applicable, the sub-contractor has the right to receive payments under the sub-contract without the statutory deduction referred to in s.69(4) of the Act.

20A.3 **Withdrawal or cancellation of tax certificate**

20A.3 .1

The sub-contractor is obliged to inform the contractor in writing if the tax certificate produced by him or referred to in the document produced by him is withdrawn or cancelled together with the date of such withdrawal or cancellation.

20A.3 .2

The contractor is obliged to inform, and the sub-contractor has the right therefore to be so informed, of any change in the positions of the contractor and/or employer under fourth recital (B) or (C).

20A.4 .1

Vouchers – I and P certificates

Where the tax certificate produced to the contractor is in one of the forms numbered 714I or 714P, the sub-contractor is obliged immediately upon receipt of any payment from which the statutory deduction has not been made to issue to the contractor a voucher in the form numbered 715.

This is largely an administrative exercise since the sub-contract is silent as regards penalties for non-production of the required voucher, but it is not unknown for contractors to withhold further payments until the voucher is produced. Any such action is contractually incorrect, but sub-contractors must be aware that failure to produce the voucher timeously may lead to a report being submitted to the Inland Revenue which could lead to cancellation of their tax certificate. In the author's experience sub-contractors do not always produce the voucher as required, i.e. 'immediately upon receipt of any payment' etc. and leave themselves open to action either by the contractor or the Inland Revenue. Such action is not in the interests of the sub-contractor and sub-contractors should, therefore, pay attention to and abide strictly by the requirements of this clause.

20A.4   .2
Vouchers – S certificates

Where the tax certificate produced to the contractor is in the form numbered 714S the sub-contractor is obliged, not later than 7 days before any payment under the sub-contract becomes due, to inform the contractor in writing of the amount to be included in such payment which represents the direct cost to the sub-contractor of materials and to give the contractor a special voucher in the form numbered 715S. The sub-contractor has the right to receive payment without any deduction provided the remainder of the payment, as indicated by the voucher, does not exceed £150 during any one week.

Again the sub-contract is silent as to penalties for non-production of the voucher and the comments in the previous clause are equally applicable here, but it is also suggested that non-production of the details required by this clause may lead to statutory deduction against all monies due under the sub-contract.

20A.4   .3

20A.5   Statutory deduction – direct cost of materials

20A.5 .1

Should the contractor be required to make the deduction referred to in the Act before a payment is due under the sub-contract, the sub-contractor has the right to be so notified in writing and the sub-contractor is obliged to state, in accordance with the time scales stated, the amount to be included in the payment which represents the direct cost to the sub-contractor of the materials used or to be used.

20A.5 .2

The sub-contractor is obliged to indemnify the contractor against any loss or expense arising from any incorrect statement provided by the sub-contractor under clause 20A.5.1.

20A.5 .3

20A.6 Correction of errors

The contractor is obliged to correct any errors or omissions in calculating or making the deduction referred to in the Act and the sub-contractor has, therefore, both a right and an obligation to abide by the provisions of this clause.

20A.7 Relation to other provisions of sub-contract

The sub-contractor is obliged to comply with the provisions of clause 20A even if it involves the sub-contractor not complying with other provisions of the sub-contract.

20A.8 Application of arbitration agreement

If any dispute or difference as to the application of clause 20A occurs, the sub-contractor is obliged to abide by the provisions of article 3 and refer such dispute or difference to resolution by arbitration in accordance with clause 38, excepting only where the act or the regulations as stated in the clause provide for some other method of resolving the dispute or difference.

20B Finance (No.2) Act 1975 – Tax deduction scheme – sub-contractor not user of a current tax certificate

Immediate attention is drawn to footnote [q] which refers in turn to the fourth recital (A). Fourth recital (A) confirms that if the sub-contractor is not a user of a current sub-contractor's tax certificate, i.e. the word 'is' is deleted, the provisions of clause 20B apply to the sub-contract and clause 20A.2 to .8 does not apply, i.e. clause 20A.1, the definition of the Act, does apply.

20B.1 Statutory deduction – direct cost of materials

20B.1 .1

The sub-contractor is obliged to advise the contractor in writing within the period specified of the amount to be included in any payment which represents the direct cost of the materials so that the appropriate deduction can be made in accordance with the Act.

Deductions are only required to be made in respect of the labour element of the sub-contract works and a statement of the material content allows the contractor to apply the correct deduction to an agreed amount which will not be later disputed.

20B.1 .2

The sub-contractor is obliged to indemnify the contractor against any loss or expense arising from any incorrect statement provided by the sub-contractor under clause 20B.1.1.

20B.1 .3

20B.2 Correction of errors

The contractor is obliged to correct any errors or omissions in calculating or making the deduction referred to in the Act and the sub-contractor has, therefore, both a right and an obligation to abide by the provisions of this clause.

20B.3 Contractor – change in regard to user of tax certificate

The contractor is obliged to inform, and the sub-contractor has the right, therefore, to be so informed, of any change in the positions of the contractor and/or employer under fourth recital (B) or (C).

20B.4 Relation to other provisions of sub-contract

The sub-contractor is obliged to comply with the provisions of clause 20B even if it involves the sub-contractor not complying with other provisions of the sub-contract.

For example, payments to the sub-contractor must be made in accordance with the provisions of clause 21 and clause 21.3 details how the amount due is to be ascertained. There is no provision within clause 21.3 for tax deductions in accordance with clause 20B and clause 20B.4, therefore, allows that deduction without being in breach of the requirements of clause 21.3.

20B.5 Sub-contractor – position where he becomes the user of a current tax certificate

The sub-contractor is obliged to notify the contractor immediately in writing if he becomes the user of a current tax certificate under the act, following which clause 20A.2 to .8 will be reinstated in substitution for clause 20B with amendments as stated.

20B.6 Application of arbitration agreement

If any dispute or difference as to the application of clause 20B occurs, the sub-contractor is obliged to abide by the provisions of article 3 and refer such dispute or difference to resolution by arbitration in accordance with clause 38, excepting only as provided in the clause.

21 Payment of sub-contractor*

21.1 First and interim payments – final payment

The sub-contractor has the right to be paid, but only in accordance with the provisions of clause 21.

*See also Housing Grants, Construction and Regeneration Act 1996, page 215.

Much has been said in the technical press and in government about unfair contract terms, most of them centring around the payment provisions – late payment, pay when paid etc. Most contractors amend the standard payment provisions, the most common amendments being to extend the payment period from the as written 17 days to 28 or even 35 days or to incorporate pay when paid provisions. These amendments do not, however, alter the principles of clause 21. In the author's experience very few sub-contractors fully understand their rights and obligations under this clause and it is suggested that many of the payment problems are self-inflicted because of this lack of understanding.

The opening statement in clause 21.1 is quite specific and unless sub-contractors do take time to acquaint themselves with their obligations under the payment clause, dispute is inevitable. It is essential that sub-contractors comply *fully* with the requirements since there are built-in safeguards in the event of failure to make payment in the manner provided in clause 21, but it may be argued that any failure by the sub-contractor to comply with his obligations may result in loss of right to payment, albeit only temporary until such time as the sub-contractor does comply fully. This can only lead to cash flow problems, dispute and possibly acrimony and breaking-off of business relationships. Failure to comply should, therefore, be avoided at all costs and *the only way to avoid it is to fully understand the obligations imposed by the sub-contract and comply with them.*

21.2   .1
Date first payment is due

Most sub-contractors ignore the statement in this clause, but it is very important as it is the key, not only to when the first payment is made, but also to the dates when all interim payments become due.

21.2   .2, .3
Date interim payments are due and made

The dates interim payments become due are fixed by the date of the first payment which they follow at intervals *not exceeding one month.*

Payment for first and interim payments are to be made *no later than 17 days* after the date they become due and the sub-contractor has the right to be so paid.

It is important to understand that there is no 'valuation due date' or any other terminology in the sub-contract; the key to payment is the *'date the first and interim payment is due'* and everything revolves around it. Thus it is important to get the first payment due date right and it is suggested that all sub-contractors should liaise with contractors and agree the best dates to accord with their (the contractor's) payment procedures.

Regular cash is the life-blood of the construction industry but, all too often, main contractors believe that cash flow only applies to them and attempt to delay payments to sub-contractors for a variety of reasons including 'computer problems', 'incorrect documentation', 'no application', 'application too late' etc. Sub-contractors, on the other hand, are constantly looking to maximise their cash at any point in time and they adopt a variety of ruses in an attempt to inflate their entitlements. Situations of delivering materials to site and immediately requesting an interim payment against them are common, as is the situation of deliberately over-booking the quantity of materials on site or work done. There has to be a compromise position which satisfies both contractor and sub-contractor, and sub-contractors should be aware of the following basic principles:

(1) Works must have *commenced* before the first payment can become due, except for off-site works if so agreed.
(2) Over-booking of materials on site or work done is so easily checked that all it does is bring the sub-contractor's credibility into question.
(3) Timing of payment due dates should be agreed having regard to the main contract procedures (see 4 below); whilst this is not written into DOM/1, it is a practical solution to problems of payment due to differing payment dates under main and sub-contracts.
(4) Most contracts using DOM/1 as the sub-contract form will be under JCT 80 as the main contract and if payment due dates coincide (and there is no reason why they should not but every reason why they should), the contractor will receive his

95

cash from the employer after 14 days (JCT 80 clause 30.1.1.1) and will pay the sub-contractor 3 days thereafter, i.e. 17 days after the payment due date. Thus there should be no reason for extended payment provisions under the sub-contract, although it is a very common occurrence.

21.3 Ascertainment of amounts due in first and interim payments

The sub-contractor has the right to have included in the first and each interim payment an amount equal to the gross valuation referred to in clause 21.4, but is obliged to allow deductions for the amounts listed under sub-clauses .1, .2 and .3.

Further mention must be made in respect of sub-clause .2 relating to discount. Cash discounts should only be allowed if payment is made in accordance with clause 21.2, i.e. the *whole* of clause 21.2. Any payment must be made within 17 days after the date when it becomes due and that due date must be within one month calculated from the date when the first payment was due. If any one of these parameters is exceeded, discount should not be deducted, i.e. if the first payment due date is more than one month after commencing works on site or if any interim payment date is more than one month thereafter or if payment is made more than 17 days after the due date. This may be better understood in graphic form (see below).

Sub-contractors should also note that discount is only to be deducted in respect of amounts referred to in clause

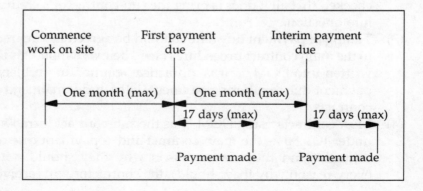

21.4.1, i.e. the total value of the sub-contract work on-site properly executed, the total value of materials and goods on-site and the total value of materials and goods off-site.

Any amounts included in respect of :

(a)   statutory obligations, notices, fees and charges (main contract clause 6)
(b)   levels and setting out of the works (main contract clause 7)
(c)   architect's instructions – clauses 17.2 and 17.3 of the main contract conditions (sub-contract clause 14.4)
(d)   disturbance of regular progress of sub-contract works – sub-contractor's claims (sub-contract clause 13.1)
(e)   payment for restoration etc. of work done under clause 8B.3 by sub-contractor (sub-contract clause 8B.4) or under clause 8C.3 by sub-contractor (sub-contract clause 8C.4)
(f)   contribution, levy and tax fluctuations (sub-contract clause 35) or labour and materials cost and tax fluctuations (sub-contract clause 36)

and any amounts deducted in respect of:

(a)   non-complying work (main contract clause 8.4.2, sub-contract clause 4.3.3.2)
(b)   architect's instructions – clauses 17.2 and 17.3 of the main contract conditions (sub-contract clause 14.4)
(c)   contribution, levy and tax fluctuations (sub-contract clause 35) or labour and materials cost and tax fluctuations (sub-contract clause 36)

as referred to in clause 21.4.2 and .3 are *not* subject to a deduction for discount.

Many main contractors seek to amend the wording of discount to 'main contractor's discount' as opposed to cash discount in the sub-contract. The intention of such an amendment is to give the main contractor the right to deduct discount irrespective of whether he pays on time or not, which is against the original intent and spirit of the sub-contract. It may be argued that the provisions of clause 21.3 do not allow for deduction of discount other than a cash discount unless clause 21.3 is itself amended. Any such amendments should be resisted and sub-contractors should insist on the cash discount provision as intended in the sub-contract, since otherwise the sub-contractor has no leverage to force payment other than suspension of work.

21.4   Gross valuation

The sub-contractor has a right to payment based on a gross valuation comprising the total of the amounts listed in clause 21.4.1 and 21.4.2 less the total amount referred to in 21.4.3 up to and including a date *not more than 7 days* before the date when the first and each interim payment is due.

Reference was made in earlier paragraphs to clause 21 of the importance of establishing the 'date when the first and each interim payment is due' and the phrase appears again here; its importance cannot be over-emphasised.

Consider, for example, a sub-contract which commences on the fifth of the month. The sub-contractor will be entitled to a payment due date no later than the fifth of the following month, with payment being made no later than the twenty second of that following month, but clause 21.1 confirms that payments to the sub-contractor are to be made in accordance with the provisions of clause 21. The gross valuation, therefore, is to be carried out no later than 7 days before the fifth of the following month, i.e. the twenty eighth of the month in which work commences if it is a thirty day month.

Sub-contractors traditionally wish to submit an application for payment, although there is no contractual requirement for them to do so, but, if that is the sub-contractor's wish, he must do so in accordance with the date requirements laid down in the sub-contract. If he fails to comply, it may be argued that he is not complying fully with clause 21 and the author does not need to spell out the consequences of that argument.

21.4   .1   .1, .2

The sub-contractor has a right to payment for materials and goods delivered to or adjacent to the works, but only to the extent that *they are reasonably, properly and not prematurely delivered* and *are adequately protected against weather and other casualties.*

Most sub-contractors believe that they have a right to payment for materials as soon as they are delivered but examination of the words in bold above clearly indicates that this is not so.

The author can quote a case from experience concerning delivery of light fittings for a hospital contract, the main contract being JCT 80 with clause 40, use of price adjustment formulae, applying. The sub-contractor saw an opportunity of making money by having all of the light fittings for the contract delivered very early in the contract, for which he believed he would receive payment as materials on site, but he would receive fluctuations on the material content of them only when they were fixed some 12 months later. They were stored in an unheated container and payment for them was withheld on the grounds that they were prematurely delivered and inadequately protected against weather and other casualties.

21.4   .1 .3
Payment for off-site goods and materials

Payment for off-site goods and materials is purely at the discretion of the architect under clause 30.3 of the main contract – 'the amount stated as due in an Interim Certificate may in the discretion of the Architect include the value of any materials or goods before delivery thereof to or adjacent to the Works' – and the sub-contractor does *not* have any automatic rights to payment therefor.

Should the architect exercise his discretion and decide to include an amount for off-site goods and materials, the sub-contractor is obliged to observe all relevant conditions set out in the main contract which have to be fulfilled before the architect is empowered to include such an amount.

Clause 30.3 sets out all of the conditions to be observed,

but three particular conditions do cause problems and are worthy of discussion.

Clause 30.3.2 confirms that 'nothing remains to be done to the materials to complete the same up to the point of their incorporation into the Works'. This is precise and means exactly what it states, e.g. structural steelwork members which have been totally fabricated and are ready for delivery to site can be included; raw steelwork which has still to be fabricated fails to comply with the requirements and cannot be included.

Clause 30.3.3 confirms that 'the materials have been and are set apart at the premises where they have been manufactured or assembled or are stored, and have been clearly and visibly marked'. Setting apart normally does not provide great difficulty, but the marking is required to show clearly the owners of the goods in question and, as such, needs to be indelible. If nothing remains to be done to the goods, the marking is unlikely to be subsequently covered and the question of how such marking can be realistically and practically carried out arises, e.g. taking the structural steelwork in the last paragraph, this may have been shot-blasted and painted and may not require further painting on site. How can it be marked in a way that will not show later? How can carpets, fittings and the like, or electrical cabinets, fans, air-handling units and the like, be marked so that the marking subsequently does not show? The author would not attempt to provide an answer to cover all situations – each situation is individual and must be considered on its merits – but clearly it is a dilemma which requires ingenuity and lateral thought to resolve.

Equally the question arises as to what extent the materials and goods should be marked. The author is aware of a case where it was agreed to pay for lift equipment as off-site materials and goods. The goods were set apart and marked, the lift motors being marked in paint on the base and the lift car, being manufactured and complete with the frame, opening gear etc. marked on the outside plywood walls. The lift manufacturer went into liquidation and when the contractor went to retrieve his lift, the lift motors and car were at the works but the frame, opening gear etc. – i.e. all the parts which were not marked and which could be easily removed – had been stripped from the car. Would it have been realistic to expect the manufacturer to have marked every composite part of the equipment, indeed would he have been prepared to do so?

Clause 30.3.9 requires that the materials and goods are insured for their full *value*. Most sub-contractors in this situation rely upon their normal works insurance, but generally this is insufficient on three counts since most standard insurance policies:

(1) only cover the *cost* of materials and goods, which may be very much less than their *value*
(2) contain an excess against each and every claim and the goods, therefore, will not be insured for their *full* value
(3) do not recognise the interests of third parties, i.e. the employer and the contractor (and the sub-contractor if sub-sub-contractors are involved).

The only answer is a separate policy geared to the particular circumstances, but that will be at extra cost and the sub-contractor will be expected to bear that extra cost since the sub-contractor will generally be the party requiring payment for off-site materials. It may, of course, be that payment is required because of delays on the main contract, and provided the sub-contract documents have been correctly prepared and notice for commencement given under clause 11 (see the articles of agreement, part 4, and clause 11.1) the sub-contractor would be entitled to give notice of delay requiring an extension of time under clause 11 and notice of loss and expense requiring reimbursement of same under clause 13.

21.4   .2, .3

21.4   .4
The sub-contractor is obliged to provide any details reasonably necessary to substantiate any statement submitted by him as to the amount of any valuation.

What is 'reasonably necessary' is undefined and open to interpretation, but it is suggested that a statement of 'To work executed – £x' is not reasonable, although such a statement is very common. If sub-contractors wish to submit a statement, and it must be repeated that there is no obligation on the sub-contractor to submit any statement, that statement should contain, as a minimum, a percentage of work executed, a detailed list of materials on-site and detailed calculations of variations, dayworks and the like. It must also be said that if a sub-contractor does not submit a statement, most contractors will not make any payment notwithstanding their obligations under the sub-contract, which are to make payment in accordance with clause 21.

21.4   .5

Property in unfixed materials and goods

21.4   .5   .1

The sub-contractor is obliged not to remove unfixed materials and goods from site unless the architect (through the contractor) and the contractor have consented in writing.

21.4   .5   .2, .3

Materials and goods become the property of the employer, where their value has been included in any interim certificate under the main contract, and the sub-contractor is obliged not to deny that such materials and goods have so become the property of the employer, or the contractor, where their value has not been so included but the contractor has paid for them.

The sub-contractor's obligation under clause 21.4.5.2 and .3 may conflict with established case law on retention of title. In any contract of supply which contains retention of title provisions, and by far the majority contain such provisions, the law would find that title in respect of goods supplied rests with the supplier until such time as the goods are paid for in full, provided that such goods are identifiable as goods in their supplied state, i.e. not fixed in position. It would appear, therefore, that until such time as the sub-contractor holds title to such goods, the sub-contactor will be unable to fulfil his obligations under the clause.

21.4 .5 .4

21.5 Retention – rules for ascertainment
.1, .2, .3

These three sub-clauses provide the rules by which retention may be deducted and retained by the contractor, and the sub-contractor is obliged to abide by them, as is the contractor. The sub-contractor does, however, have the right to release of one moiety of the retention fund where the sub-contract works have reached practical completion as defined in clause 14.

Practical completion of the sub-contract works is *not* dependent upon practical completion of the works; the trigger for practical completion of the sub-contract works is the notice required under clause 14.1 and deemed practical completion under clause 14.2 can only apply in the event of dissent to the sub-contractor's notice. Hence it is important if sub-contractors wish to avail themselves of their right under clause 21.5.2 that they should have given the notice required under clause 14.1.

Sub-contractors have a right to release of the final moiety of the retention fund upon the issue of the certificate of completion of making good defects under clause 17.4 of the main contract or a certificate under clause 18.1.3 of the main contract relating to partial possession.

Main contractors never, in the author's experience, pay retention monies when the certificate(s) as aforesaid are issued and if the sub-contractor does not ask, he is unlikely ever to see his retention money. The sub-contractor will, however, be aware of the contract date for completion and the defects liability period from the information issued at tender stage and that contained in the appendix to the articles of agreement, part 1 section B. Unless, therefore, the contractor has been awarded an extension of time, the sub-contractor will be aware of when the defects liability period should end and, allowing a reasonable period for clearance of defects, suggested as three months, a reasonable estimate of the likely date for the issue of the certificate of making good defects can be made. As a starting point to monitoring final release of the retention fund, and it will need to be monitored regularly, the contractor should be requested to confirm the practical completion date of the works and to provide details of any partial possessions to which clause 18 of the main contract applies. Adding the defects liability period to the dates provided will give the sub-contractor a date, after which he should regularly ask if the certificate under clause 17.4 or 18.1.3, if applicable, has been issued. If payment of the final moiety of the retention fund is not made by the contractor in accordance with clause 21.5.3 (and it very rarely is) any cash discount under clause 21.3.2 should be disallowed.

21.6   Right of sub-contractor to suspend execution of sub-contract works*

In the event of non-payment as provided in clause 21, the sub-contractor has the right to suspend execution of the sub-contract works, but only provided that the sub-contractor shall have given the contractor written notice of same and then only provided that default shall have continued for 7 days after giving notice. Thus the sub-contractor's right is dependent upon his obligation to give notice if he wishes to invoke his right.

Few sub-contractors use this clause in the sub-contract to overcome late payments by contractors, but it is a powerful tool. If invoked, the sub-contractor's actions are not 'deemed a failure to proceed with the Sub-Contract Works' and any costs to the sub-contractor arising from the suspension would be recoverable from the contractor under clause 13.1 as an act, omission or default of the contractor. Clearly such an action will cause a problem between contractor and sub-contractor, but it is possible that if more sub-contractors were prepared to use their right under this clause contractors might become better paymasters.

21.7   .1

Final adjustment of sub-contract sum

Where clause 15.1 applies (lump sum contract), the sub-contractor is obliged to send to the contractor, *not later than 4 months after practical completion of the sub-contract works* (see

*See also Housing Grants, Construction and Regeneration Act 1996, page 215.

clause 14), all documents necessary for the purpose of the adjustment of the sub-contract sum.

This clause does not require the sub-contractor to carry out the adjustments – only to provide the documents necessary – but, again in the author's experience, contractors never adjust the sub-contract sum of their own volition and wait for individual sub-contractors to submit their own 'final accounts'. Submission of a final account clearly establishes what the sub-contractor feels his entitlements are and, at the very least, establishes a starting point for both parties to agree the final adjustments due. It is suggested that any final account submitted should follow the format in clause 21.7.2; following such a format will ensure that all matters are considered.

21.7   .2
Items included in sub-contract sum

21.8   .1
Computation of ascertained final sub-contract sum

Where clause 15.2 applies (subject to complete re-measurement and valuation), the sub-contractor is obliged to send to the contractor, *not later than 4 months after practical completion of the sub-contract works* (see clause 14), all documents necessary for the purpose of computing the ascertained final sub-contract sum.

Again this clause does not require the sub-contractor to compute the sum – only to provide the documents necessary – but, again in the author's experience, contractors never compute the final sub-contract sum of their own volition and wait for individual sub-contractors to submit their own 'final accounts'. Submission of a final account clearly establishes what the sub-contractor feels his entitlements are and, at the very least, establishes a starting point for both parties to agree the final ascertained sum. It is suggested that any final account submitted should follow the format in clause 21.8.2; following such a format will ensure that all matters are considered.

21.8   .2
Items included in ascertained final sub-contract sum

21.9 **.1,.2**

Amount due in final payment

Date of final payment

These clauses contain three rights for the sub-contractor.

The sub-contractor has the right to be paid the amount calculated in accordance with clause 21.7 or 21.8 as appropriate, less only discount and previous payments as stated, and the final payment becomes due not later than 7 days after the date of issue of the final certificate issued by the architect under clause 30.8 of the main contract conditions. Before the date that final payment becomes due, the sub-contractor has the right to be notified by the contractor in writing by registered post or recorded delivery of the amount of the final payment to be made, and finally the sub-contractor has the right to be paid the final payment within 28 days of the date that it becomes due.

This is all very clear and straightforward, except that sub-contractors generally are not aware of the date of issue of the final certificate under clause 30.8 of the main contract conditions and the contractor is not under any obligation to tell the sub-contractor of the date of its issue. The sub-contractor's clear and straightforward right suddenly becomes clouded and, as in clause 21.5, monitoring becomes necessary, using as guides the date of practical completion of the works, the defects liability period, the date of the issue of the certificate of making good defects under clause 17.4 and the requirements of clause 30.8 of the main contract conditions. Clause 30.8 of the main contract states that the final certificate shall be issued by the architect not later than two months after whichever of the listed events last occurs, and it is possible to approximately calculate this likely date, assuming a 12-month defects liability period, as shown in the following table.

Each column in the table represents the total period from practical completion of the works to the issue of the final certificate in respect of each of the events listed in clause 30.8:

Event	1 (months)	2 (months)	3 (months)
Practical completion of the works	0	0	0
Defects liability period in main contract	12	12	
Make good defects/issue of certificate of making good defects		3	
Issue documents for adjustment of contract sum (clause 30.6.1.1 of main contract)			6
Ascertainment and statement (clause 30.6.1.2.1 and .2 of main contract)			3
Issue of final certificate	2	2	2
Total period from practical completion of the works	14	17	11

(1) the end of the defects liability period
(2) the issue of the certificate of making good defects
(3) the ascertainment and statement required by clause 30.6 of the main contract.

It can be seen that the earliest date for the issue of the final certificate on a contract with a 12 months defects liability period is 14 months from practical completion of the works, but column 2 is more likely to apply to most contracts. Sub-contractors should, however, carefully check the defects liability period since six months for building work and 12 months for mechanical and electrical installations and landscaping only are quite common and, whilst the period for mechanical and electrical etc. will determine the date for issue of the final certificate, a certificate of making good defects for building work will determine the date for final release of retention monies for that part (see clause 21.5).

21.10 Effect of final payment

107

22 Benefits under main contract

The sub-contractor has the right to have the contractor obtain for him any rights or benefits of the main contract but only *so far as the same are applicable to the sub-contract works* and only at the request and at the cost, *if any*, of the sub-contractor.

Many contractors argue that obtaining extensions of time or recovering loss and/or expense is a main contract benefit and they seek to recover the cost of 'passing on' the sub-contractor' claims. This not so; the sub-contractor has a right to extensions of time and recovery of loss and/or expense under the provisions of the sub-contract (clauses 11 and 13) and any attempt by contractors to raise cost claims for this should be strongly resisted. This clause is limited only to those rights and benefits which are *not* available under the sub-contract and, for all practical purposes, is unlikely ever to be invoked (see also clause 5.1).

23 Contractor's right to set-off*

23.1 Agreed amounts – amounts awarded in arbitration or litigation

The sub-contractor is obliged to allow the contractor to deduct from any money otherwise due under the sub-contract any amount *agreed by the sub-contractor* or *finally awarded in arbitration or litigation and which arises out of or under the sub-contract.*

The sub-contractor's obligations and, per se, the contractor's rights are severely limited under this clause. An amount can *only* be deducted if:

(a) it is agreed by the sub-contractor, or
(b) it is awarded in arbitration or litigation, and
(c) it arises out of or under the sub-contract.

Thus awards arising out of or under other sub-contracts *cannot*, as of right, be deducted, but they may be if agreed by the sub-contractor.

*See also Housing Grants, Construction and Regeneration Act 1996, page 215.

23.2   Amounts not agreed

.1, .2

The contractor has a right to set-off against any monies due under the sub-contract, but only where the contractor has a claim for loss and/or expense and/or damage suffered or incurred by reason of any breach of, or failure to observe the provisions of, the sub-contract by the sub-contractor. In order for the contractor to invoke his right under this clause he must comply strictly with the requirements of clause 23.2.2 and:

(a)   have quantified the set-off *in detail and with reasonable accuracy*

(b)   have sent the details of the quantification to the sub-contractor

(c)   have given the sub-contractor notice in writing of his intention to set-off the amount quantified, and

(d)   have given the notice not less than 3 days before the date upon which the payment from which the contractor intends to make the set-off becomes due under clause 21.2.1 or clause 21.2.2.

Note particularly the requirement under (d) that the notice relates to the date the payment becomes *due* and *not the date the payment is due to be made.*

Sub-contractors should also note that any claim of set-off by the contractor will typically arise under the provisions of clause 4.5 and/or clause 12.2 and/or clause 13.4 and all of these clauses require the contractor to have given notice in writing. It follows, therefore, that unless *all* of the relevant notices have been given in accordance with the provisions of the sub-contract, i.e. under the clause giving rise to the loss and/or expense and/or damage and under clause 23.2, the contractor has no right of set-off.

23.3   Further rights of set-off etc.

23.4   Exclusion of implied terms on set-off

24 Contractor's claims not agreed by the sub-contractor – appointment of adjudicator*

24.1 Sub-contractor disagrees with set-off – action by sub-contractor – action by adjudicator

24.1 .1, .2

If the sub-contractor disagrees with the amount specified in the written notice issued by the contractor under clause 23.2.2, he has the right, within 14 days of receipt by him of the notice, to send to the contractor by registered post or recorded delivery a written statement setting out the reasons for such disagreement and particulars of any counterclaim against the contractor. At the same time the sub-contractor is obliged to give notice of arbitration *and* request action by the adjudicator in accordance with the right given under clause 24.1.2 *and* send to the adjudicator by registered post or recorded delivery a copy of his (the sub-contractor's) statement, the written notice of the contractor and brief particulars of the sub-contract.

Sub-contractors should note that they are not obliged to invoke the adjudication provisions within clause 24 – clause 24.1.1, line 3, states 'the Sub-Contractor may' – but if they wish to invoke them they must comply with all of the provisions and they must do so within 14 days of receipt of the written notice of the contractor. If sub-contractors fail to act within the time scale stated, and it is very short if they wish to produce a counterclaim which has to be quantified in detail and with reasonable accuracy, they will lose their right to adjudication, but not their right to arbitration.

The sub-contractor, under the provisions of clause 24.1.2, finally has the right to request the adjudicator named in the appendix part 8 to act as the adjudicator to decide those matters referable to him under clause 24.

24.2 Contractor's written statement

The contractor has the right, within 14 days from receipt of the sub-contractor's statement of counterclaim, to send to

*See also Housing Grants, Construction and Regeneration Act 1996, page 215.

the adjudicator a written statement setting out brief particulars of his defence to the counterclaim. The contractor is not obliged to set out a defence but, if he wishes to, he must do it within the time scale specified otherwise he will lose his right under this clause.

The contractor is obliged to send to the sub-contractor, and the sub-contractor, therefore, has the right to receive, a copy of the written statement of defence.

Sub-contractors should be aware that the contractor's right is limited to a defence to the counterclaim only and any attempt to introduce additional matters supporting his claim should be resisted.

24.3   Decision of Adjudicator

.1, .2, .3
Both the contractor and the sub-contractor are obliged to be bound by the adjudicator's decision but only until the matters upon which he has given his decision have been settled by agreement or determined by an arbitrator or the court. Both the contractor and the sub-contractor have the right to be notified of the adjudicator's decision.

Thus any decision of the adjudicator is an interim decision only on the contractor's right of set-off and for no other purpose; the adjudication process does not replace arbitration or litigation and any dispute under the sub-contract will have to be resolved either by negotiation and agreement or by arbitration under clause 38.

Adjudication under this clause should not be confused with arbitration. The sub-contractor has a right under the sub-contract to present a case to a third party, the adjudicator, stating why set-off should not be allowed; the contractor has a right to respond. The adjudicator makes a decision based upon the written statements given by the sub-contractor and contractor only; no other evidence is allowed and no appearence or questioning before the adjudicator takes place. The decision of the adjudicator is an interim decision only and is binding only until agreement between the parties or until an award in arbitration.

Arbitration, on the other hand, is carried out under the provisions of the Arbitration Act and the parties appear before a judge, the arbitrator, who has the right to examine all evidence presented and to request further evidence if he is of the opinion that it helps to establish the facts and enables him to reach the correct decision. The parties have the right to be represented, have the right to present all evidence in support of their case and can be questioned by opposing counsel. The Arbitration Act, being an act of parliament, has the full backing of the law in the enforcement of any awards made.

24.4   Implementation of adjudicator's decision

.1, .2

The adjudicator's decision will determine how the amount of set-off notified by the contractor is to be dealt with (see clause 24.3.1) and, where that decision requires the contractor to pay an amount to the sub-contractor, the sub-contractor has the right to be paid such amount immediately upon receipt of the decision of the adjudicator.

24.5   Trustee-Stakeholder

.1, .2

Where the adjudicator's decision requires the contractor to deposit an amount with a trustee-stakeholder, the amount has to be held in trust in a deposit account with any interest earned added to the sum deposited and the sub-contractor has, therefore, a right to the interest, the primary deposit being the sub-contractor's money withheld by the contractor, but pending the outcome in arbitration or subsequent negotiations. The sub-contractor is obliged to notify the trustee-stakeholder of the name and address of the adjudicator and arbitrator referred to in clause 24.

24.6   Adjudicator's decision – power of arbitrator

24.7   Further sums – set-off and counterclaims

Any action of set-off under clause 23.2 and/or counterclaim under clause 24.1.1 is not finite and both contractor and sub-contractor have the right to similar actions as applicable when further sums become due to the sub-contractor.

24.8 Adjudicator's fee – charges of trustee-stakeholder

The sub-contractor is obliged to pay the adjudicator's fee and any charges of the trustee-stakeholder, but the arbitrator in his final award will settle ultimate responsibility for those payments.

25 Right of access of contractor and architect

The sub-contractor is obliged to give the contractor and the architect and all persons duly authorised by either of them access at all reasonable times to any work being prepared for or utilised in the sub-contract works, but not further or otherwise, particularly where the sub-contractor wishes to protect any proprietary rights.

Clearly where the sub-contractor has patented an operation or owns rights to an operation, he is under no obligation to divulge the 'trade secret' behind his operation; nor is he obliged to allow access to work which does not form part of the sub-contract works, but sub-contractors should consider whether this could be accommodated, if requested, as part of a marketing exercise.

26 Assignment – sub-letting

26.1 Sub-contractor not to assign without consent

If the sub-contractor wishes to assign the sub-contract he is obliged to obtain the written consent of the contractor before so doing, but sub-contractors should be aware that the contractor is *not* under any obligation to agree to such assignment, whether reasonable or otherwise.

Failure to comply with the requirements of this clause is a breach of the terms of the sub-contract which could lead to determination of the employment of the sub-contractor under clause 29.1.4.

26.2 Sub-contractor not to sub-let without consent

If the sub-contractor wishes to sub-let any portion of the sub-contract works he is obliged to obtain the written consent of the contractor before so doing and such consent is not to be unreasonably withheld, but the sub-contractor remains wholly responsible for carrying out and completing the whole of the sub-contract works notwithstanding that a portion or portions may be sub-let.

Very few sub-contractors comply with the requirements of this clause but sub-contractors should be aware that failure to comply is a breach of the terms of the sub-contract for which they become liable and which could lead to determination of the employment of the sub-contractor under clause 29.1.4. Consent to sub-letting is required in respect of the act of sub-letting only; the sub-contractor is under no obligation to divulge the name of the sub-sub-contractor, although divulging the name can be a help to sub-contractors in some circumstances.

Consider, for example, a situation where a portion of the sub-contract works is to be sub-let to a sub-sub-contractor of whom the contractor has prior knowledge and doubt as to his ability to carry out and complete the sub-let portion. Consent to sub-letting under those circumstances could be reasonably withheld, but if the sub-contractor fails to notify his intentions, the contractor is denied his rights and the sub-contractor would be liable for all costs arising; those costs may not have arisen had the sub-contractor obtained the benefit of the contractor's experience or the sub-contractor may have been able to obtain assurances such that the doubts arising from the previous experiences of the contractor could have been allayed.

27 Attendance

27.1 .1
Attendance

The sub-contractor has the right to be provided by the contractor with the attendance items listed in the clause free of charge.

The facilities to be provided free of charge are for the purpose of the sub-contract works and many sub-contractors argue that their offices, storage sheds etc., which are essential for the carrying out of the sub-contract works, are as much a part of the sub-contract works as the materials incorporated therein and the facilities should, therefore, be provided to the accommodation. Clause 27.3 (see later) clarifies that temporary services to workshops, sheds or other temporary buildings are to be provided at the sub-contractor's expense. The listed items can be open to interpretation, e.g. what are reasonable hoisting facilities? The answer is very much dependent upon the nature of the sub-contract works and should be clarified at tender stage.

If, for example, heavy items of equipment requiring crane off-loading are required – such as air-handling plant, generators and the like – sub-contractors should clarify within their tender whether crane off-loading for that equipment is or is not included. Similar provisions would apply for materials delivered on pallets, e.g. windows and doors. The provision of space for the storage of materials is precisely the provision of space only and the contractor is not under any obligation to provide secure space. Sub-contractors should, therefore, be aware that materials on site remain totally their responsibility (see clause 8A or clause 8B or clause 8C as applicable) and it is their responsibility to provide such security for their materials as they consider reasonable; the watching provided by the contractor under this clause is not intended to abrogate the sub-contractor's responsibilities for security of his own materials.

27.1 .2
Scaffolding for sub-contract works

The sub-contractor has the right to be provided by the contractor with all necessary scaffolding and scaffold boards for work over 11 feet high free of charge. The sub-contractor is obliged to provide his own scaffolding and scaffold boards for work 11 feet high or under.

27.1 **.3**
Clearance of rubbish

The sub-contractor is obliged to clear away all rubbish arising from the execution of the sub-contract works and, upon practical completion of the sub-contract works, to clear up properly and leave the works and all areas specified clean and tidy.

This obligation should not be ignored by sub-contractors but, alas, it often is. Current legislation places more and more onus upon employers to take responsibility for the safety of their own and other employees, and keeping the workplace clean and tidy is of paramount importance in this regard. Most contractors will specify a skip as being the place provided and that is reasonable provided the access to the skip is reasonable, e.g. on a multi-storey development it would not be reasonable to expect sub-contractors to carry rubbish up and down ladders – that is in itself dangerous – and sub-contractors could expect to be provided with access to a chute at each floor level. Equally on a large site, one skip in one corner would be unrealistic.

27.2 **Particular items of attendance**

The sub-contractor has the right to be provided by the contractor with the attendance items detailed in the appendix part 9 free of charge.

Any particular items required by the sub-contractor, such as the craneage debated under clause 27.1.1 above, should be listed in the appendix part 9. The sub-contractor has no other right to attendances other than those listed in clause 27 and the appendix part 9 and it is, therefore, important that adequate provision be made in the appendix at sub-contract placement; anything omitted at that stage cannot, without extreme difficulty, be added later.

27.3 **Workshops, etc. of sub-contractor**

The sub-contractor is obliged to provide, erect, maintain, move and remove all necessary workshops, sheds or other temporary buildings at his own expense, but he has the right to all reasonable facilities from the contractor for such erection, maintenance etc.

As confirmed in the commentary under clause 27.1.1 above and stated in clause 27.3, it is the sub-contractor's responsibility to provide the temporary services to his own workshops etc. Those services, dependent upon the nature and complexity of the sub-contract, will comprise water, electricity, telephone and, in rare cases, drainage. Separate water and electricity connections to the contractor and sub-contractors on one site, if allowed by the utility companies, would prove cost prohibitive and sub-contractors will be bound to purchase their utility supplies from the contractor. High rates of supply for these services are not unknown and sub-contractors should attempt to agree the level and cost of supply at sub-contract placement, although there is no reason why these facilities cannot be agreed as contractor supplied attendances and listed in the appendix part 9.

27.4 Use of erected scaffolding of the contractor or sub-contractor

The sub-contractor has both an obligation to allow the use of and a right to use any erected scaffolding belonging to or provided by the sub-contractor or the contractor as appropriate *while it remains so erected upon the site.*

This clause does not supersede the obligations of the contractor or the sub-contractor under clause 27.1.1; any scaffolding to be supplied by the contractor or the sub-contractor as appropriate under the terms of the sub-contract clause 27.1.1 must be so provided. This clause relates to other standing scaffolding which one party or the other may make use of in the carrying out of the works, e.g. if the sub-contractor is carrying out work to a wall under 11 feet high and has erected scaffold for that purpose, the contractor or others on the site have the right to use that scaffold for works to be carried out on the same wall. The obligation and/or right is only to the extent that the use of the scaffolding is for the purpose of the works; no party has an obligation to keep scaffolding erected for as long as other persons wish to use it and no obligation exists as to the fitness, condition or suitability of the said scaffolding.

28 Contractor and sub-contractor not to make wrongful use of or interfere with the property of the other

The contractor and the sub-contractor are obliged not to use wrongfully or interfere with the plant, ways, scaffolding, temporary works, appliances or other property belonging to or provided by the other, nor to be guilty of any infringement of act of parliament or bye-law etc., but the obligation cannot prejudice or limit the rights of the parties in the carrying out of their statutory or contractual duties.

This clause protects one party's equipment etc. on site from misuse or interference by the other, but only to the extent that there is no statutory or contractual obligation. Consider, for example, the situation where it has been agreed in the sub-contract that, whilst the contractor will provide and erect scaffolding over 11 feet high, the sub-contractor will move scaffold boards as required to suit his work; clearly the sub-contractor has a contractual duty to move those boards and his actions would not place him in breach of this clause.

29 Determination of the employment of the sub-contractor by the contractor

29.1 Default by sub-contractor

In the event of default for one of the reasons listed in the clause, the contractor has the right to issue a notice specifying the default, but he is not obliged to do so. If the contractor does choose to issue a notice under the provisions of this clause, that notice must be sent to the sub-contractor by registered post or recorded delivery. Only after continuance of the default by the sub-contractor for *10 days after receipt of the notice* or any subsequent *repeat of the default* does the contractor have the right to determine the employment of the sub-contractor.

29.2   Sub-contractor becoming bankrupt etc.

In the event of the bankruptcy or insolvency or winding up or receivership of the sub-contractor, the contractor has the right to determine the employment of the sub-contractor, but again he is not obliged to do so. In the event that the contractor does choose to determine the employment of the sub-contractor, he is obliged to issue a written notice to that effect and the employment of the sub-contractor is immediately determined upon the issue of the written notice.

29.3   Corruption

29.4   Contractor and sub-contractor – rights and duties

.1
The contractor has a right, and the sub-contractor is there-fore obliged to allow him, to use all temporary buildings, plant etc. as listed and to purchase all materials and goods necessary for the carrying out and completion of the sub-contract works.

29.4   .2
Except where determination occurs by reason of bankruptcy etc. (under clause 29.2), the sub-contractor is obliged, if so required by the contractor within 14 days of the date of determination, to assign to the contractor without payment the benefit of any agreement for the supply of materials or goods and/or the execution of any work for the purposes of the sub-contract.

Note that the obligation on the sub-contractor is dependent upon action by the contractor *within 14 days of the date of determination*. If the contractor fails to act within the time scale stated he loses his right under this clause. If the contractor fails to act or chooses not to take the benefit of any assignment available and makes his own arrangements to carry out and complete the sub-contract works, the contractor will seek to recover those costs under the provisions of clause 29.5 (see later), but if the sub-contractor can demonstrate that the costs paid by the contractor were more than the sub-contractor would have paid, the sub-contractor could reasonably claim that that cost did not represent any direct loss arising out of the determination.

29.4 .3
The sub-contractor is obliged, as and when required by a direction of the contractor, to remove from the works any temporary buildings etc. as listed.

If the sub-contractor fails to comply with the directions of the contractor, the contractor has the right to sell the property of the sub-contractor in the manner specified in the clause, but note that the sub-contractor's obligation and the contractor's right are subject to the *issue of a direction of the contractor* (see clause 4.2). If no direction is issued, the contractor has no right to dispose of the sub-contractor's property and if he does so dispose without the issue of a direction, the sub-contractor would have a claim against the contractor for the value of the property, notwithstanding any proceeds actually realised by the contractor.

29.5

The sub-contractor is obliged to allow or pay to the contractor the amount of any direct loss and/or damage caused to the contractor by the determination. Until after completion of the sub-contract works, the sub-contractor has no right to any further payment under any provision of the sub-contract, but upon such completion he has the right to apply for the value of any work executed or goods and materials supplied, to the extent that their value has not been included in previous interim payments, and the contractor is obliged to pay, less only any discount specified in the appendix part 7 and the amount of any direct loss and/or damage.

Note that the sub-contractor only has an obligation to reimburse *direct* loss and/or damage and only if *caused by the determination*. Many contractors view determination as giving them the right to retain any further payments due to the sub-contractor, but this clearly is not so. The action under the provisions of clause 29 determines the *employment* of the sub-contractor only; the sub-contract remains a valid, legal document and both contractor and sub-contractor are still bound by all of the conditions contained therein.

Clause 29.4 and clause 29.5 set out the respective rights and duties of the contractor and the sub-contractor following determination; their respective rights and duties prior to determination are as set out in the Sub-Contract DOM/1. Consider, for example, the situation with retention; clause 29.5 makes no mention of the retention fund and it is quite clear that, in the event of determination and following completion of the sub-contract works, the contractor has no right to deduct further retention monies on any payment due under this clause. Any retention monies deducted prior to determination under the provisions of clause 21.5 will still be bound by the provisions of clause 21.5 after determination.

30 Determination of employment under the sub-contract by the sub-contractor

30.1 Acts etc. giving grounds for determination of employment by sub-contractor

In the event of default for one of the reasons listed in the clause, the sub-contractor has the right to issue a notice specifying the default, but he is not obliged to do so. If the sub-contractor does choose to issue a notice under the provisions of this clause, he is obliged to send that notice to the contractor by registered post or recorded delivery. Only after continuance of the default by the contractor for *10 days after receipt of the notice* or any subsequent *repeat of the default* does the sub-contractor have the right to determine his employment.

In the event of suspension of the sub-contract works under clause 21.6, the sub-contractor has no right to issue a notice under clause 30.1.1.3 until 10 days after the date of commencement of the suspension, and his right to determination under this provision will not occur until a minimum of 31 days after any payment was due to be made, i.e.:

Period from written notice under clause 21.6 to suspension	7 days
Period from suspension to issue of notice under clause 30.1.1.3	10 days
Period from notice under clause 30.1.1.3 to determination	10 days
Postage allowance	4 days

During this period the sub-contractor is obliged to continue to progress the sub-contract works in accordance with the provisions of the sub-contract and he should not countenance withdrawal or reduction of resources, as that could lead to notice(s) under the provisions of clause 29 of the sub-contract.

30.2   Determination of employment by sub-contractor – rights and duties of contractor and sub-contractor

The respective rights and duties under clause 30.2 are without prejudice to the accrued rights or remedies of either party or to any liability under clause 6, i.e. this clause does not contain the sole rights or remedies under the sub-contract and the rights or remedies accrued under other provisions of the sub-contract are equally valid.

30.2   .1

The sub-contractor is obliged to remove from the site all his temporary buildings, plant, tools etc. as listed.

30.2   .2

The sub-contractor has the right to be paid by the contractor for all the items listed in clause 30.2.2.1 to .5 inclusive.

Note that no reference is made to the time within which payment is to be made, but, as only the employment of the sub-contractor is determined and the provisions of the sub-contract still exist, payment must be made in accordance with the time provisions within clause 21.

31   Determination of the contractor's employment under the main contract

If the employment of the contractor is determined under the provisions of the main contract conditions, the employment of the sub-contractor under the sub-contract shall thereupon also determine and the sub-contractor has both a right and an obligation to abide by the provisions of clause 30.2 of the sub-contract.

32 Number not used

33 Strikes – loss and/or expense

33.1 Strikes etc. – position of contractor and sub-contractor

.1, .2, .3
The sub-contractor has no right to recovery of, nor any obligation to reimburse, any loss and/or expense resulting from strike action howsoever arising.

The sub-contractor has the right of access to the site, which the contractor is obliged to take all reasonably practical steps to keep open and available for use, and the sub-contractor is obliged to take all reasonably practical steps to continue with the sub-contract works.

33.2 Preservation of rights of contractor and sub-contractor

The sub-contractor's rights and obligations under other provisions of the sub-contract are preserved and nothing in clause 33 can affect those rights and obligations.

34 Choice of fluctuation provisions

Immediate attention is dawn to footnote [f] which confirms that clause 36 should be used where the parties have agreed to allow the labour and materials cost and tax fluctuations to which clauses 36.1 to .3 refer or clause 37 should be used where the parties have agreed that the sub-contract sum or the tender sum shall be adjusted by the formula method under the sub-contract/works contract formula rules.

34.1 Choice of fluctuation provisions

The sub-contractor is obliged to use whichever alternative is stated to apply in the appendix part 10.

The sub-contractor has no right to recovery of any other increased cost except as provided within clause 35 or clause 36 or clause 37 as applicable, although if an unforeseen and unrecoverable increase occurs as a result of an action which would give rise to reimbursement under the provisions of clause 13 of the sub-contract, that increase could quite properly be recovered under those provisions.

34.2   Clause 35 to apply if no other provision stated

The sub-contractor is obliged to abide by the provisions of clause 35 (contribution, levy and tax fluctuations) where neither clause 36 nor clause 37 is stated in the appendix part 10.

35   Contribution, levy and tax fluctuations

35.1   Deemed calculation of sub-contract sum or tender sum – types and rates of contribution etc.

The sub-contractor is obliged to adjust the sub-contract sum or tender sum if one or more of the events specified occurs.

35.1   .1

Clause 35.1.1 confirms the basis of the prices contained in the sub-contract sum or tender sum at the base date, which is the date inserted in the appendix part 11, and under current legislation refers to the employer's National Insurance contributions.

35.1   .2
Increases or decreases in rates of contribution etc. – payment or allowance

If any of the tender rates is increased or decreased or if a tender type ceases to be payable or if a new type of contribution, levy or tax becomes payable after the base date, then the sub-contractor has the right to be paid any increase in or new type of, and is obliged to allow any decrease in or cessation of, such contribution, levy or tax as the case may be.

Any recovery or allowance under this clause is limited to

the workpeople engaged upon or in connection with the sub-contract works as listed in the clause and it is essential, in order to effect recovery under this clause, that adequate records be kept to demonstrate that workpeople were so engaged upon or in connection with those sub-contract works.

35.1 .3, .4
Persons employed on site other than workpeople

The right and obligation of the sub-contractor under clause 35.1.2 are extended to other persons employed by the sub-contractor and engaged upon or in connection with the sub-contract works, but who are not within the definition of 'workpeople' contained in clause 35.6.3; the payment or allowance in respect of such people is as for a craftsman under clause 35.1.2, but subject to the restrictions and definitions contained in clause 35.1.4.

35.1 .5, .6, .7
Refunds and premiums

If any of the tender rates is increased or decreased or if a tender type ceases to be payable or if a new type of refund of contribution, levy or tax becomes receivable or if a new type of premium becomes receivable after the base date, then the sub-contractor has the right or obligation, as the case may be, to be paid or to allow the difference between what the sub-contractor actually receives or will receive and what he would have received had the alteration, cessation or new type of refund or premium not become effective.

35.1 .8
Contracted-out employment

Where the sub-contractor pays contributions in respect of workpeople whose employment is 'contracted out', the sub-contractor is obliged, for the purpose of recovery or allowance under clause 35.1, to be deemed to pay contributions as if that employment were not contracted out.

35.1 .9
Meaning of contribution etc.

35.2 Materials – duties and taxes

The sub-contractor is obliged to adjust the sub-contract sum or tender sum if one or more of the events specified occurs.

35.2 .1

Clause 35.2.1 confirms the basis of the prices contained in the sub-contract sum or tender sum at the base date in respect of materials, goods etc. as listed and attached to the appendix part 11, including, but only where specifically agreed by the contractor and sub-contractor, fuels.

35.2 .2

If any of the tender rates is increased or decreased or if a tender type ceases to be payable or if a new type of duty or tax (other then value added tax) becomes payable after the base date, then the sub-contractor has the right or obligation, as the case may be, to be paid or to allow the difference between what the sub-contractor actually pays and what he would have paid had the alteration, cessation or imposition not occurred.

35.3 Fluctuations – work sub-let to sub-sub-contractors

35.3 .1

Sub-let work – incorporation of provisions to like effect

If the sub-contractor sub-lets any portion of the sub-contract works he is obliged to incorporate in the sub-sub-contract provisions to have the same effect as the provisions of clause 35 (excluding clause 35.3) including the percentage stated in the appendix part 11 pursuant to clause 35.7.

The intention behind the clause is that any sub-sub-contractor should have the same rights or obligations to recovery or allowance of contribution, levy and tax fluctuations as the sub-contractor under clause 35, but there are no rules as to how those provisions should be incorporated into any sub-sub-contract. The same provisions or similar provisions or re-written provisions could be incorporated; any possible solution is acceptable as long as it complies with the requirements of clause 35.3 and passes the same rights and obligations to the sub-sub-contractor.

Note that the operation of this clause is subject to clause

26.2 (sub-contractor not to sub-let without consent) and failure to comply with the requirements of clause 26.2 would lead to loss of any rights to recover increases under clause 35. Note also that clause 35.3 is excluded from the sub-sub-contract and the sub-contractor, therefore, has no right to recovery of increases applicable to sub-sub-sub-contracts and sub-sub-sub-letting should, as a general rule, not be contemplated.

35.3 .2

Sub-let works – fluctuations – payment to or allowance by the sub-contractor

If the price payable under the sub-sub-contract changes by reason of the operation of the incorporated provisions, then the sub-contractor has the right or obligation, as the case may be, to be paid or to allow any increase or decrease arising therefrom.

35.4
to .6

Provisions relating to clause 35

35.4 .1

Written notice by sub-contractor

The sub-contractor is obliged to give written notice to the contractor of the occurrence of any of the events referred to in the listed provisions, relating to an increase in or decrease of or cessation of or new type of contribution, levy or tax, including any refund or premium, in connection with the employment of people and/or duties and taxes in respect of materials and/or similar provisions in connection with sub-let work.

35.4 .2

Timing and effect of written notices

The sub-contractor is obliged to give any notice required by clause 35.4.1 within a reasonable time after the occurrence of the event to which the notice relates.

The giving of a written notice is a condition precedent to any payment being made to the sub-contractor and it is, therefore, vital that relevant written notices should be given, but

127

it is interesting that the sub-contract is silent as regards notice being a condition precedent to any allowances which the sub-contractor is obliged to give. It must be assumed that the contractor is entitled to any decreases irrespective of the notice required by clause 35.4.1 being given.

35.4   .3
Agreement – contractor and sub-contractor

35.4   .4
Fluctuations added to or deducted from the sub-contract sum or included in the calculation of the ascertained final sub-contract sum

Clause 35.4.4 should be read in conjunction with clause 3 and provides that any adjustment under clause 35.1 and .2 or clause 35.3 is to be included in interim certificates and the sub-contractor has the right to be paid therefor.

There appears to be an anomaly in the wording of clause 35.4.4; line 2 refers to clause 35.1 *and* .2 *or* clause 35.3 which could be interpreted to indicate that only payment or allowances in respect of *either* workpeople and materials directly employed by the sub-contractor *or* sub-sub-contractors are covered by the provisions of clause 35.4.4, but not both. Such an interpretation cannot have been the intent behind the wording when the clause was drafted, since such an intent would have envisaged total sub-letting of the sub-contract works, which would, in the author's view, be unrealistic.

35.4   .5
Evidence and computations by sub-contractor

The sub-contractor is obliged to provide such evidence and computations as the contractor may reasonably require to enable the amount to be paid or allowed to be ascertained, including a certificate signed by or on behalf of the sub-contractor each week certifying the validity of the evidence.

35.4   .6
No alteration to sub-contractor's profit

The sub-contractor has the right to the profit included in the sub-contract sum or ascertained final sub-contract sum as appropriate and is under no obligation to adjust that profit for any addition or deduction arising from the operation of clause 35.4.4.

35.4   Position where sub-contractor in default over completion

35.4   .7

The sub-contractor has no right to adjustment of the amount payable to or allowed by the sub-contractor if the event referred to in clause 35.4.1 occurs after the date of failure by the sub-contractor to complete as notified by the contractor under clause 12.1, and the sub-contractor is obliged to bear the cost or benefit of such adjustments himself.

The operation of clause 35.4.7 is subject to the provisions of clause 35.4.8 (see below) and, clearly, the issuing of the notice by the contractor under clause 12.1; if no notice is issued the provisions of this clause cannot be applied.

35.4   .8   .1

If the printed text of clauses 11.2 to .10 is amended in any way, the provisions of clause 35.4.7 cannot be applied and the sub-contractor has the right to recovery or allowance for all adjustments arising from any of the events listed in clause 35.4.1.

35.4   .8   .2

If the contractor has failed to give a decision in writing in respect of revisions to the period or periods for completion of the sub-contract works for which the sub-contractor has issued a written notice under clause 11.2, the provisions of clause 35.4.7 cannot be applied and the sub-contractor has the right to recovery or allowance for all adjustments arising from any of the events listed in clause 35.4.1.

The contractor must have given his decision *in writing* and in respect of *every* written notice – even a failure to respond to one notice out of a number will lead to the non-application of clause 35.4.7 – and it is vital that sub-contractors comply in all respects with their obligations under clause 11.2.

35.5 **Work etc. to which clauses 35.1 to .3 not applicable**

35.6 **Definitions for use with clause 35**

35.7 **Percentage addition to fluctuation payment or allowances**

35.7 .1

The sub-contractor has the right or obligation, as the case may be, to add to any amount due or any amount to be allowed the percentage stated in the appendix part 11.

36 Labour and materials cost and tax fluctuations

36.1 **Deemed calculation of sub-contract sum or tender sum – rates of wages etc.**

The sub-contractor is obliged to adjust the sub-contract sum or tender sum if one or more of the events specified occurs.

36.1 .1

Clause 36.1.1 confirms the basis of the prices contained in the sub-contract sum or tender sum at the base date, which is the date inserted in the appendix part 11, as the rates of wages etc. and any contribution, levy or tax payable by the sub-contractor in respect of workpeople engaged upon or in connection with the sub-contract works as listed.

36.1 .2

Increases or decreases in rates of wages etc. – payment or allowance

If any of the wage rates etc. is increased or decreased by promulgation after the base date, then the sub-contractor has the right to be paid any such increase and is obliged to allow any such decrease, as the case may be, together with any consequential increase or decrease in the cost of insurances and any consequential increase or decrease in the cost of any contribution, levy or tax.

Any recovery or allowance under this clause is limited to the workpeople engaged upon or in connection with the sub-contract works as listed in clause 36.1.1 and it is essential, in order to effect such recovery, that adequate records be kept to demonstrate that workpeople were so engaged upon or in connection with those sub-contract works. Sub-contractors should also note that only increases or decreases arising from alterations in the rules decisions or agreements of the relevant wage fixing body (see clause 36.1.1.3, .4 and .5) are subject to adjustment.

36.1   .3, .4
Persons employed on site other than workpeople

The right and obligation of the sub-contractor under clause 36.1.2 are extended to other persons employed by the sub-contractor and engaged upon or in connection with the sub-contract works but who are not within the definition of 'workpeople' contained in clause 36.7.3; the payment or allowance in respect of such people is as for a craftsman under clause 36.1.2, but subject to the restrictions and definitions contained in clause 36.1.4.

36.1   .5, .6
Workpeople wage-fixing body – reimbursement of fares etc.

If any of the transport charges as referred to in the list attached to the appendix part 12 is increased or decreased after the base date, or if the reimbursement of any fares is increased or decreased by promulgation of the relevant wage fixing body after the base date, then the sub-contractor has the right to be paid any such increase and is obliged to allow any such decrease, as the case may be.

36.2   **Contributions, levies and taxes**

36.2   .1
Clause 36.2.1 confirms the basis of the prices contained in the sub-contract sum or tender sum at the base date and under current legislation refers to the employer's National Insurance contributions.

36.2   .2
If any of the tender rates is increased or decreased, or if a tender type ceases to be payable, or if a new type of contribution, levy or tax becomes payable after the base date, then the sub-contractor has the right to be paid any increase

in or new type of, and is obliged to allow any decrease in or cessation of, such contribution, levy or tax as the case may be.

Any recovery or allowance under this clause is limited to the workpeople engaged upon or in connection with the sub-contract works as referred to in clause 36.1.1.1 and 36.1.1.2 and it is essential, in order to effect recovery under this clause, that adequate records be kept to demonstrate that workpeople were so engaged upon or in connection with those sub-contract works.

36.2 .3

The right and obligation of the sub-contractor under clause 36.2.2 are extended to other persons employed by the sub-contractor and engaged upon or in connection with the sub-contract works, but who are not within the definition of 'workpeople' contained in clause 36.7.3; the payment or allowance in respect of such people is as for a craftsman under clause 36.2.2, but subject to the restrictions and definitions contained in clause 36.1.4.

36.2 .5, .6

If any of the tender rates is increased or decreased, or if a tender type ceases to be payable, or if a new type of refund of contribution, levy or tax becomes receivable, or if a new type of premium becomes receivable after the base date, then the sub-contractor has the right or obligation as the case may be to be paid or to allow the difference between what the sub-contractor actually receives or will receive and what he would have received had the alteration, cessation or new type of refund or premium not become effective.

36.2 .7

Where the sub-contractor pays contributions in respect of workpeople whose employment is 'contracted out', the sub-contractor is obliged, for the purpose of recovery or allowance under clause 36.2, to be deemed to pay contributions as if that employment were not contracted out.

36.2 .8

36.3   Materials, goods, electricity and fuels

The sub-contractor is obliged to adjust the sub-contract sum or tender sum if one or more of the events specified occurs.

36.3   .1

Clause 36.3.1 confirms the basis of the prices contained in the sub-contract sum or tender sum at the base date in respect of materials, goods etc. as listed and attached to the appendix part 12, including, but only where specifically agreed by the contractor and sub-contractor, fuels.

36.3   .2

If any of the market prices are increased or decreased, then the sub-contractor has the right or obligation as the case may be to be paid or to allow the difference between the basic price and the market price payable by the sub-contractor.

The sub-contractor's right or obligation under this clause is limited only to those materials, goods, electricity and fuels listed in the appendix part 12 – any elements not included are at the sole cost and risk of the sub-contractor – and only to any increase or decrease in *market* prices. Any local variations in prices do not constitute changes in market prices and are not recoverable or allowable.

36.4   Fluctuations – work sub-let to sub-sub-contractors

36.4   .1
Sub-let work – incorporation of provisions to like effect

If the sub-contractor sub-lets any portion of the sub-contract works he is obliged to incorporate in the sub-sub-contract provisions to have the same effect as the provisions of clause 36 (excluding clause 36.4) including the percentage stated in the appendix part 12 pursuant to clause 36.8.

The intention behind the clause is that any sub-sub-contractor should have the same rights or obligations to recovery or allowance of labour and materials costs and tax fluctuations as the sub-contractor under clause 36, but there are no rules as to how those provisions should be incorporated into any sub-sub-contract. The same provisions or similar provisions or re-written provisions could be incorporated; any possible solution is acceptable as long as it complies with the requirements of clause 36.4 and passes the same rights and obligations to the sub-sub-contractor.

Note that the operation of this clause is subject to clause 26.2 (sub-contractor not to sub-let without consent) and failure to comply with the requirements of clause 26.2 would lead to loss of any rights to recover increases under clause 36. Note also that clause 36.4 is excluded from the sub-sub-contract and the sub-contractor, therefore, has no right to recovery of increases applicable to sub-sub-sub-contracts, and sub-sub-sub-letting should, as a general rule, not be contemplated.

36.4 .2
Sub-let works – fluctuations – payment to or allowance by the sub-contractor

If the price payable under the sub-sub-contract changes by reason of the operation of the incorporated provisions, then the sub-contractor has the right or obligation, as the case may be, to be paid or to allow any increase or decrease arising therefrom.

36.5 to .7
Provisions relating to clause 36

36.5 .1
Written notice by sub-contractor

The sub-contractor is obliged to give written notice to the contractor of the occurrence of any of the events referred to in the provisions listed in the clause and relating to an increase in or decrease of wage rates or other emoluments, fares etc. in connection with the employment of people, and/or an increase in or decrease of or cessation of or new type of contribution, levy or tax, including any refund or premium, in connection with the employment of people and/or an increase in or decrease of the market prices, duties and taxes in respect of materials and/or similar provisions in connection with sub-let work.

36.5 .2
Timing and effect of written notices

The sub-contractor is obliged to give any notice required by clause 36.5.1 within a reasonable time after the occurrence of the event to which the notice relates.

The giving of a written notice is a condition precedent to any payment being made to the sub-contractor and it is, therefore, vital that relevant written notices should be given, but it is interesting that the sub-contract is silent as regards notice being a condition precedent to any allowances which the sub-contractor is obliged to give. It must be assumed that the contractor is entitled to any decreases irrespective of the notice required by clause 36.5.1 being given.

36.5 .3
Agreement – contractor and sub-contractor

36.5 .4
Fluctuations added to or deducted from the sub-contract sum or included in the calculation of the ascertained final sub-contract sum

Clause 36.5.4 should be read in conjunction with clause 3 and provides that any adjustment under clause 36.1 to .3 or clause 36.4 is to be included in interim certificates and the sub-contractor has the right to be paid therefor.

There appears to be an anomaly in the wording of clause 36.5.4; line 2 refers to clause 36.1 to .3 *or* clause 36.4 which could be interpreted to indicate that only payment or allowances in respect of *either* workpeople and materials directly employed by the sub-contractor *or* sub-sub-contractors are covered by the provisions of clause 36.5.4, but not both. Such an interpretation cannot have been the intent behind the wording when the clause was drafted, since such an intent would have envisaged total sub-letting of the sub-contract works, which would, in the author's view, be unrealistic.

36.5 .5
Evidence and computations by sub-contractor

The sub-contractor is obliged to provide such evidence and

computations as the contractor may reasonably require to enable the amount to be paid or allowed to be ascertained, including a certificate signed by or on behalf of the sub-contractor each week certifying the validity of the evidence.

36.5 .6

No alteration to sub-contractor's profit

The sub-contractor has the right to the profit included in the sub-contract sum or ascertained final sub-contract sum as appropriate and is under no obligation to adjust that profit for any addition or deduction arising from the operation of clause 36.5.4.

36.5 **Position where sub-contractor in default over completion**

36.5 .7

The sub-contractor has no right to adjustment of the amount payable to or allowed by the sub-contractor if the event referred to in clause 36.5.1 occurs after the date of failure by the sub-contractor to complete, as notified by the contractor under clause 12.1, and the sub-contractor is obliged to bear the cost or benefit of such adjustments himself.

The operation of clause 36.5.7 is subject to the provisions of clause 36.5.8 (see below) and, clearly, the issuing of the notice by the contractor under clause 12.1; if no notice is issued the provisions of this clause cannot be applied.

36.5 .8 .1

If the printed text of clauses 11.2 to .10 is amended in any way, the provisions of clause 36.5.7 cannot be applied and the sub-contractor has the right to recovery or allowance for all adjustments arising from any of the events listed in clause 36.5.1.

36.5 .8 .2

If the contractor has failed to give a decision in writing in respect of revisions to the period or periods for completion of the sub-contract works for which the sub-contractor has issued a written notice under clause 11.2, the provisions of clause 36.5.7 cannot be applied and the sub-contractor has the right to recovery or allowance for all adjustments arising from any of the events listed in clause 36.5.1.

The contractor must have given his decision *in writing* and in respect of *every* written notice – even a failure to respond to one notice out of a number will lead to the non-application of clause 36.5.7 – and it is vital that sub-contractors comply in all respects with their obligations under clause 11.2.

36.6 **Work etc. to which clauses 36.1 to .4 not applicable**

36.7 **Definitions for use with clause 36**

36.8 **Percentage addition to fluctuation payment or allowances**

36.8 .1
The sub-contractor has the right or obligation, as the case may be, to add to any amount due or any amount to be allowed the percentage stated in the appendix part 12.

37 Formula adjustment

37.1 **Use of sub-contract/works contract formula rules**

The sub-contractor has both a right and an obligation to adjust the sub-contract sum or amounts ascertained under clause 17 in accordance with the provisions of clause 37 and the sub-contract/works contract formula rules identified in the appendix part 13.

37.2 **Miscellaneous provisions on applying formula adjustment**

37.2 .1, .2

37.2 .3
The sub-contractor has a right and an obligation to carry out valuations for the purpose of calculating formula adjustments under clause 37.

37.3   Miscellaneous provisions cont.

37.3   .1

The sub-contractor has the right to have any adjustments arising from the operation of clause 37 effected in all interim payments to which clause 21 applies and the contractor is specifically obliged to ensure that such effect is carried out.

37.3   .2

The sub-contractor has the right to make to the contractor any representations on the value of the work to which the formula adjustment applies.

37.3   .3, .4

Non-adjustable elements only apply where the main contract is let on the Standard Form of Building Contract, Local Authorities Edition – see footnote [k] to the appendix part 13.

37.4   Fluctuations – articles manufactured outside the UK

For any article manufactured outside the UK to which rule 4(ii) of the formula rules applies, the sub-contractor is obliged to insert in the appendix part 13 the market price of the article in sterling current at the base date. If the market price of the article changes after the base date, the sub-contractor has the right or obligation, as the case may be, to be paid or to allow the net increase or decrease in the market price current when the article is bought; market price in this context is deemed to include any duty or tax (other than value added tax as noted).

37.5   Power to agree – contractor and sub-contractor

If the contractor agrees with the sub-contractor any alteration to the methods and procedures as noted, the sub-contractor is obliged to abide by those altered methods and procedures.

The operation of this clause is subject to the proviso commencing on line 6 – that no alterations to the methods and procedures shall be agreed unless it is reasonably expected that the amount of adjustment will be the same or approximately the same as that ascertained in accordance with the rules. It does not confer any right on the contractor or sub-contractor for wholesale departure from the methods and procedures in the applicable formula rules.

37.6   Position where monthly bulletins are delayed

37.6   .1, .2, .3

In the event of delay in, or cessation of, the publication of the monthly bulletin, the sub-contractor has a right and an obligation to adjustment of the sub-contract sum or amounts ascertained under clause 17 on a fair and reasonable basis, but only during the period of delay, and both the contractor and sub-contractor are obliged to operate such parts of clause 37 and the applicable formula rules as will enable any adjustments to be readily calculated upon recommencement of publication. In the event of recommencement of publication the provisions of clause 37 and the applicable formula rules apply and the sub-contractor is obliged to comply therewith.

37.7   Formula adjustment – failure to complete

37.7   .1

The sub-contractor's right to adjustment of the amount payable to or allowed by the sub-contractor, after the date of failure by the sub-contractor to complete (as notified by the contractor under clause 12.1), is calculated by reference to the index numbers applicable to the valuation period in which the date or dates of expiry of the completion or completions of the sub-contract works occur, and the sub-contractor is obliged to bear the cost or benefit of adjustments thereafter himself.

The operation of clause 37.7.1 is subject to the provisions of sub-clauses .2 and .3 and, clearly, the issuing of the notice by the contractor under clause 12.1; if no notice is issued the provisions of this clause cannot be applied.

37.7   .2

If the contractor has failed to give a decision in writing in respect of revisions to the period or periods for completion of the sub-contract works for which the sub-contractor has issued a written notice under clause 11.2, the provisions of clause 37.7.1 cannot be applied and the sub-contractor has the right to recovery or allowance, as the case may be, for all adjustments arising from the operation of the provisions of clause 37.

The contractor must have given his decision *in writing* and in respect of *every* written notice – even a failure to respond to one notice out of a number will lead to the non-application of clause 37.7.1 – and it is vital that sub-contractors comply in all respects with their obligations under clause 11.2.

37.7   .3

If the printed text of clauses 11.2 to .10 is amended in any way, the provisions of clause 37.7.1 cannot be applied and the sub-contractor has the right to recovery or allowance, as the case may be, for all adjustments arising from the operation of the provisions of clause 37.

38   Settlement of disputes – arbitration

38.1

If the sub-contractor requires any dispute or difference to be referred to arbitration, then the sub-contractor is obliged to give written notice to the contractor to such effect.

38.2   .1, .2, .3

If the dispute or difference to be referred to arbitration raises issues which are substantially the same as or connected with issues raised in a related dispute between parties as listed, and if the related dispute has already been referred for determination to an arbitrator, the contractor and the sub-contractor have agreed and are obliged to refer the dispute or difference under the sub-contract to the arbitrator appointed to determine the related dispute, save as provided in sub-clauses .2 and .3 to clause 38.2.

38.3

38.4

38.5

38.6

38.7

The sub-contractor has the right to appeal to the High Court on any question of law arising out of an award made in arbitration under the agreement or arising in the course of the reference.

38.8

38.9

38.10

The sub-contractor is obliged to abide by the JCT Arbitration Rules and footnote [v] confirms that these rules contain stricter time limits than those prescribed by other rules or observed in practice.

The JCT rules not only contain strict time constraints but also limit what can be introduced into the arbitration in an effort to streamline the process and curb costs. As a general rule the legal profession do not like the limitations imposed by these rules and opposing counsel will normally seek to agree at the preliminary hearing not to be bound by the rules, but open the arbitration to all of the normal legal processes – discovery, document bundles etc. In the author's view such agreements merely open the door for ever-increasing costs and, having entered into an agreement to abide by the JCT arbitration rules, both parties to the contract should indeed be bound by them, although, also in the author's view, the plaintiff will have more to gain from departing from the rules than the defendant.

The Codes of Practice

To the extent that operation of the codes of practice is a contractual requirement, both parties to the sub-contract have a legal obligation to operate and abide by the contents. They are, however, codes of practice, not conditions of contract, and the only legal obligation on the parties is that the various matters listed within the code should be considered prior to taking action under the conditions.

Code of Practice 'A': referred to in clause 4.3.2.1

Clause 4.3.2.1 refers to the issuing of directions requiring the removal from site or rectification of non-complying work and requires the contractor to consult with the sub-contractor and have regard to this code of practice *prior to issuing* the said directions.

1.

The intent of the code is to assist the contractor in deciding whether directions under clause 4.3.2.1 should be issued. Many contractors believe that they have an inalienable right to issue directions and that those directions cannot be challenged, but unless the contractor complies with the code he has no rights at all and the sub-contractor may challenge any direction issued without consultation.

2.

The contractor and the sub-contractor are obliged to endeavour to agree whether non-complying work is to be removed, but, in any event, the contractor is obliged to consider the criteria listed in the clause and the sub-contractor, therefore, has the right to such consideration.

Code of Practice 'B': referred to in clause 4.3.2.3

Clause 4.3.2.3 refers to the issuing of directions requiring the opening up for inspection or testing of non-complying work and requires the contractor to have regard to this code of practice in issuing the said directions.

1.

The intent of the code is to assist in the fair and reasonable operation of the requirements of clause 4.3.2.1. Again many contractors believe that they have an inalienable right to issue directions and that those directions cannot be challenged, but unless the contractor complies with the code he has no rights at all and the sub-contractor may challenge any direction issued without regard to the code.

2.

The contractor and the sub-contractor are obliged to endeavour to agree the amount and method of opening up or testing, but, in any event, the contractor is obliged to consider the criteria listed and the sub-contractor, therefore, has the right to such consideration.

The Amendments

Amendment 4 – issued September 1989
Incorporation of 'Schedules of Work' and 'Contract Sum Analysis'

Amendment 4, which is only for use where Amendment 3 to JCT 80 applies and only to the Without Quantities version of the form, requires certain deletions and alterations to be made to the provisions of Sub-Contract DOM/1. The required alterations are set out in detail within the amendment, but the deletions only are contained in the 'IMPORTANT NOTICE' at the foot of page 1, which confirms that, being a lump sum contract with no provision for measurement and valuation, article 2.2 of Sub-Contract DOM/1 will need to be deleted in every case where amendment 4 applies and that related provisions in clause 15.2 (Price for sub-contract works: tender sum – ascertained final sub-contract sum), clause 17 (Valuation of all work comprising the sub-contract works) and clause 18 (Bills of quantities – standard method of measurement), together with any reference elsewhere in the conditions, e.g. clause 30, to the 'Tender Sum' or 'Ascertained Final Sub-Contract Sum', will similarly be inapplicable.

The amendments to the recitals in the articles of agreement are quite major and they are, therefore, produced in full below, following a similar format to those earlier in this book.

The Articles of Agreement

The articles open with provision for inserting the date and the names and registered addresses of the parties to the contract, i.e. 'the Contractor' and 'the Sub-Contractor'. There then follow six recitals or statements covering:

(1) The contractor desires to have work executed as referred to in the appendix.

(2) A. The sub-contractor has priced the numbered documents listed in the appendix, or

B. the sub-contractor has stated the sum he will require for carrying out the sub-contract works and has supplied the contractor with a priced sub-contract document on which that sum is based.

Whichever of the alternatives A or B is applicable is dependent upon the alternatives in the main contract as footnote [1.a].

(3) The numbered documents referred to in the appendix, part 2, together with the priced sub-contract documents have been signed by the contractor and the sub-contractor and annexed to the articles of agreement.

(4) The sub-contract works are to be carried out as part of work carried out by the contractor under a contract with the employer and provision is made in the fourth recital to insert the name of the employer.

(5) The sub-contractor has had reasonable opportunity of inspecting all of the provisions of the main contract.

Note here that the sub-contractor has a right to inspect '*all* of the provisions of the Main Contract, or a copy thereof, except the detailed prices of the Contractor included in the Priced Documents referred to in the Appendix, Part 1, Section A' and the sub-contractor must be given a reasonable opportunity of carrying out that inspection.

The emphasis given to 'all' is the author's and the sub-contractor should ensure that he does so inspect before entering into contract. Any later claims of 'not being aware' will undoubtedly fail because of the statement in this recital and the later article 1.1. Note also particular comments regarding enquiry documents in Part 1 of this book.

(6) The relevant tax status of sub-contractor, contractor and employer under the provisions of the Finance (No.2) Act 1975 and which of clauses 20A or 20B is to apply under the sub-contract.

The alterations to the articles and the appendix give credence to 'the Priced Sub-Contract Document' (as defined in the second recital B.) and it is not proposed to make any further comments here other than to say that they neither give additional rights to nor impose additional obligations on the sub-contractor.

Similarly, the alterations to the conditions give credence as aforesaid and remove references to 'Contract Bills'. Again additional comment is largely superfluous – that which has already been written in this book is equally applicable, even with the introduction of amendment 4 – save as to clause 16.3 and this clause is produced in full below so that it can be easily understood in its revised format.

16.3 Valuation rules

16.3 .1

The sub-contractor has both a right and an obligation to have work measured and valued in accordance with the rules, but only to the extent that the valuation relates to additional or substituted work. There follow sub-clauses .1 and .3 (.2 is not used with amendment 4) which set out the rules to be adopted for valuation under the particular conditions appertaining, all of which contain both a right and an obligation to abide by those rules.

16.3 .1 .1

This on the face of it is clear but sub-contractors should note that all elements must be satisfied in applying the rule: the work must be of similar character – if it is not, then this rule does *not* apply – *and* a fair allowance must be made for any change in the conditions under which the work is carried out *and* a fair allowance must be made for any significant change in the quantity of such work from that included in the sub-contract documents.

16.3 .1 .3

The next logical step after sub-clause 1. In this case, if the work is *not* of similar character to work set out in the sub-contract documents, the work shall be valued at fair rates and prices.

16.3 .2

The sub-contractor has both a right and an obligation to have omitted work measured and valued in accordance with this rule, but only to the extent that the valuation relates to *the omission of work set out in the priced sub-contract documents.*

Varied work which is subsequently omitted is not governed by this rule. There can be circumstances where, subsequent to preparation of the priced sub-contract documents, drawings have been amended prior to work starting on site and then work omitted. The quantity surveyor may seek to measure and value work omitted from the latest drawings in accordance with this rule; clearly this is incorrect and only that included in the sub-contract documents can be omitted.

16.3   .3
Clauses 16.3.1 and 16.3.2 give a right to and place an obligation on sub-contractors to use and abide by those rules; the two sub-clauses to clause 16.3.3 (clause 16.3.3.1 is not used with amendment 4), which apply to all valuations under clauses 16.3.1 and 16.3.2, are no different. Both parties to the sub-contract must abide by these rules.

16.3   .3   .2
It should be noted that percentage additions or lump sum adjustments apply to all valuations under clause 16.3.1 and 16.3.2, *including work valued at fair rates and prices under clause 16.3.1.3,* and are not restricted to those items valued in accordance with clauses 16.3.1.1, i.e. valuations consistent with the relevant rates, prices or amounts for such work in the priced sub-contract documents.

16.3   .3   .3
This is often a difficult rule to apply since generally sub-contractors do not price preliminaries separately. Compliance with clause 16.3.1.1 should not cause a problem since valuations consistent with rates, prices or amounts in the documents may well be inflated by the amount of preliminaries built into the original bid and, whilst this may not be totally satisfactory, it does give the sub-contractor a general recovery of preliminaries across all variations which may well be sufficient to cover any extra costs.

Valuation under clause 16.3.1.3 can prove difficult where there is no breakdown of the preliminaries in the original bid, but, irrespective of a provided breakdown or otherwise, sub-contractors have the right to have preliminaries considered in any valuation of a variation.

16.3 .4

Note the wording; this clause relates *only* to work which *cannot be valued by measurement*, notwithstanding that quantities do not form a part of the sub-contract documents. There is a logical progression to the way in which the rules in clause 16.3 are set out, but often sub-contractors will turn to this clause immediately as a way of ensuring that all costs are covered. *This is wrong.* Sub-contractors and contractors must follow the logical progression within clause 16.3 and use the rule which first applies to the circumstances surrounding the variation. The basic question is, 'Is it measurable?' If the answer is yes, clause 16.3.1 or clause 16.3.2 will apply; if the answer is no, clause 16.3.4 will apply.

16.3 .4 .1, .2

There is a tendency for 'specialist trades' to believe that they fall within the province of clause 16.3.4.2, no matter what their trade is. Reference to the wording of clause 16.3.4.2 and footnote [j] clearly shows that this is not true; the only trades falling within the rule in clause 16.3.4.2 are those in which the Royal Institution of Chartered Surveyors and the appropriate body representing the employers in that trade have agreed and issued a definition of prime cost of daywork. Footnote [j] confirms those bodies, and thereby the trades they represent, as being the Electrical Contractors Association, the Electrical Contractors Association of Scotland and the Heating and Ventilating Contractors Association. All other 'specialist trades' are bound by the rules in clause 16.3.4.1.

The sub-contractor has an obligation to comply with *all* of the requirements of the proviso if he considers that valuation of work executed should be in accordance with the rules in clause 16.3.4.

All too often sub-contractors fail either in timeous presentation of records for verification or in presentation of incomplete records, which they subsequently try and recover in final account negotiations. The sub-contract unfortunately makes no provision for action in the event of failure to abide by the requirements of the proviso, nor is there any established precedent, but it is suggested that, unless the sub-contractor complies with the requirements in full, the contractor has no liability to value, and subsequently make payment, in accordance with clause 16.3.4.

Amendment 7 – issued September 1989
Standard Method of Measurement 7th Edition: Bills of Quantities prepared in accordance with SMM 7

Amendment 7, which is only for use where amendment 7 to JCT 80 applies, requires certain deletions and alterations to be made to the provisions of Sub-Contract DOM/1. The required alterations, which give credence to quantities being prepared in accordance with the rules contained in the *Standard Method of Measurement* 7th edition, are set out in detail within the amendment but are further discussed below.

Articles of agreement

1.3

The wording is altered to incorporate amendment 7.

Sub-contract conditions

1.3   Definitions

Additional definitions are included in respect of 'Approximate Quantity' and 'provisional sum', together with footnote [n.1], arising from the provisions of General Rules 10.1 to 10.6 of SMM 7.

11.10   Relevant events

As stated earlier, clause 11.10 is merely a list of events

which may give rise to an extension of time under clause 11 and do not place any obligations on or give any rights to the sub-contractor. The sub-contractor's right to an extension of time arises under clause 11.3 which refers to clause 11.10 and any alterations, therefore, to the text of clause 11.10 affect the sub-contractor's rights under clause 11.3.

11.10 .5 .1

Under clause 11.10.5.1, any right to an extension of time under clause 11.3 for compliance with architect's instructions issued under main contract clause 13.3 (Instructions on provisional sums) excludes instructions regarding the expenditure of provisional sums for *defined* work, i.e. defined as required by General Rule 10.3 of SMM 7. The author has seen numerous bills of quantities which purport to contain 'defined' provisional sums but which fail to comply with the requirements of General Rule 10.3 and, under those circumstances, clause 18.1.2 (see later) will require the description of the provisional sum to be corrected so that it does comply with the requirements. No allowance for programming, planning and pricing preliminaries as required by General Rule 10.4 of SMM 7 should be included for any provisional sum which does not fully comply with General Rule 10.3, notwithstanding that it may be described in the bills of quantities as a 'Defined Provisional Sum'.

11.10 .6

Late receipt of instructions is clarified as including those for or in regard to the expenditure of provisional sums.

11.10 .15

Clause 11.10.15 is a totally new event and arises from General Rule 10.1 of SMM 7. Rule 10.1 provides that where work can be described and given in items in accordance with the rules but the quantity of work cannot be accurately determined, an estimate of the quantity shall be given and identified as an approximate quantity. Clause 11.10.15 provides for situations where the approximate quantity in the bills of quantities was not a reasonably accurate forecast of the quantity of work required.

13.2   Relevance of certain extensions of sub-contract time

The text of clause 13.2 is amended to include the new relevant event 11.10.15, but the sub-contractor's basic right under this clause, as detailed earlier, is unchanged.

13.3   Relevant matters

As stated earlier, these sub-clauses merely list those matters which can give rise to the recovery of direct loss and/or expense referred to in clause 13.1 and contain no additional rights or obligations over and above those listed in clauses 13.1 and 13.2.

13.3   .1

Late receipt of instructions is clarified as including those for or in regard to the expenditure of provisional sums.

13.3   .7

Under clause 13.3.7, any right to the recovery of direct loss and/or expense referred to in clause 13.1 for compliance with architect's instructions issued under main contract clause 13.3 (Instructions on provisional sums) is clarified to exclude instructions regarding the expenditure of provisional sums for *defined* work, i.e. defined as required by General Rule 10.3 of SMM 7 (see comments on defined provisional sums under clause 11.10 of amendment 7).

13.3   .8

Clause 13.3.8 is a totally new matter and arises from General Rule 10.1 of SMM 7. Rule 10.1 provides that where work can be described and given in items in accordance with the rules but the quantity of work cannot be accurately determined, an estimate of the quantity shall be given and identified as an approximate quantity. Clause 13.3.8 provides for situations where the approximate quantity in the bills of quantities was not a reasonably accurate forecast of the quantity of work required.

16.1   Valuation

The sub-contractor has the right to have all variations and all work executed as described valued and to have all work executed for which an approximate quantity is included in

152

the bills of quantities measured and valued, but only where clause 15.1 applies. That valuation shall, unless otherwise agreed by both parties, be made in accordance with the provisions of clause 16 and the sub-contractor has both a right and an obligation to abide by those rules.

The sub-contractor does not have a right to depart from the rules; only in the event of agreement by *both* parties can departure from the rules be envisaged.

The valuation rules embodied in clause 16.3 and clause 17.3 are produced in full below so that they can be easily understood in their revised format.

16.3   Valuation rules

16.3   .1

The sub-contractor has both a right and an obligation to have work measured and valued in accordance with the rules, but only to the extent that the valuation relates to additional or substituted work *which can be properly valued by measurement* or to the execution of work for which an approximate quantity is included in the bills of quantities, where those bills are included in the numbered documents.

Measurement in this context, as confirmed in the later clause 16.3.3.1, relates to measurement following the same principles as those used in the preparation of the bills of quantities. If particular additional or substituted work cannot be measured in accordance with those principles, then the provisions of clause 16.3.1 cannot apply and the sub-contractor must look to the other rules to determine which one should apply.

There follow sub-clauses .1, .2, .3, .4 and .5 which set out the rules to be adopted for valuation under the particular conditions appertaining, all of which contain both a right and an obligation to abide by those rules.

16.3   .1   .1

This on the face of it is clear but sub-contractors should note that this rule only applies to *additional or substituted* work and that all elements must be satisfied for the rule to apply. The work must be of similar character *and*

153

executed under similar conditions *and* not significantly change the quantity. If any one element does not apply then this rule does not apply.

16.3 .1 .2

This again only applies to *additional or substituted* work and is the next logical step after sub-clause .1, but again all required elements must be satisfied: it must be of similar character but *not* executed under similar conditions *and/or* significantly change the quantity. Two elements only need be satisfied here; if they are not then this rule cannot be used.

16.3 .1 .3

This again only applies to *additional or substituted* work and is the next logical step after sub-clause 2 – the whole of clause 16.3 is logical in the way that it is set out such that once all parameters are satisfied, that is the rule to apply. In this case only one element has to be satisfied, i.e. that the work is not of similar character.

Provided the opening to clause 16.3.1 is satisfied, i.e. that the work can be properly valued by measurement, one of the sub-clauses .1, .2 or .3 must contain the rule that is to apply to the valuation of additional or substituted work.

16.3 .1 .4

This rule only applies to any *approximate quantity* included in the bills of quantities where the approximate quantity is a reasonably accurate forecast of the quantity of work required.

16.3 .1 .5

Again this rule only applies to any *approximate quantity* included in the bills of quantities, but where the approximate quantity is *not* a reasonably accurate forecast of the quantity of work required. The rule requires that the valuation shall include a fair allowance for any difference in quantity.

The proviso to clause 16.3.1 confirms that the rules in clause 16.3.1.4 and clause 16.3.1.5 only apply to the extent that the work has not been altered or modified other than in quantity. If the work is not executed under similar conditions or is not of similar character, the principles established in clause 16.3.1.2 or clause 16.3.1.3 apply.

16.3 .2

The sub-contractor has both a right and an obligation to have omitted work measured and valued in accordance with this rule, but only to the extent that the valuation relates to *the omission of work set out in bills of quantities and/or other documents.*

Varied work which is subsequently omitted is not governed by this rule. The author is aware of circumstances where, subsequent to preparation of bills of quantities, drawings have been amended prior to work starting on site and then work omitted. The quantity surveyor has sought to measure and value work omitted from the latest drawings in accordance with this rule; clearly this is incorrect and only that measured in the bills of quantities can be omitted.

16.3 .3

Clauses 16.3.1 and 16.3.2 give a right to and place an obligation on sub-contractors to use and abide by those rules; the three sub-clauses to clause 16.3.3, which apply to all valuations under clauses 16.3.1 and 16.3.2, are no different; both parties to the sub-contract must abide by these rules.

16.3 .3 .1

Reference was made to this clause in 16.3.1 above. Unless the work can be properly valued by measurement, the provisions of clause 16.3.1 cannot apply and this clause sets out that those measurements shall be in accordance with the same principles as those governing the preparation of the bills of quantities comprised in the sub-contract documents. Neither party to the sub-contract can depart from those measurement rules.

16.3 .1 .2

It should be noted that percentage additions or lump sum adjustments apply to all valuations under clause 16.3.1 and 16.3.2, *including work valued at fair rates and prices under clause 16.3.1.3,* and are not restricted to those items valued in accordance with clauses 16.3.1.1 and

16.3.3.2, i.e. at rates set out in the bills of quantities or pro rata thereto.

16.3 .3 .3

This is often a difficult rule to apply since generally sub-contractors do not, and indeed are often not given the opportunity to, price preliminaries separately. Compliance with clauses 16.3.1.1 and 16.3.1.2 should not cause a problem since the bill rates will be inflated by the amount of preliminaries built into the original bid and, whilst this may not be totally satisfactory, it does give the sub-contractor a general recovery of preliminaries across all variations which should be sufficient to cover any extra costs.

Valuation under clause 16.3.1.3 can prove difficult where there is no breakdown of the preliminaries in the original bid, but, irrespective of a provided breakdown or otherwise, sub-contractors have the right to have preliminaries considered in any valuation of a variation, excepting only valuation of work following compliance with a direction of the contractor regarding the expenditure of a provisional sum for defined work.

16.3 .4

Note the wording; this clause relates *only* to work which *cannot be valued by measurement*. It must be repeated that there is a logical progression to the way in which the rules in clause 16.3 are set out, but often sub-contractors will turn to this clause immediately as a way of ensuring that all costs are covered. *This is wrong.* Sub-contractors and contractors must follow the logical progression within clause 16.3 and use the rule which first applies to the circumstances surrounding the variation. The basic question is, 'Is it measurable?'. If the answer is yes, clause 16.3.1 or clause 16.3.2 will apply; if the answer is no, clause 16.3.4 will apply.

16.3 .4 .1, .2

There is a tendency for 'specialist trades' to believe that they fall within the province of clause 16.3.4.2, no matter what their trade is. Reference to the wording of clause 16.3.4.2 and footnote [j] clearly shows that this is not true; the only trades falling within the rule in clause 16.3.4.2 are those in which the Royal Institution of Chartered Surveyors and the appropriate body representing the employers in that trade have agreed and issued a definition of prime cost of daywork. Footnote [j] confirms those bodies, and thereby the trades they represent, as being the Electrical Contractors Association, the Electrical Contractors Association of Scotland and the Heating and Ventilating Contractors Association. All other 'specialist trades' are bound by the rules in clause 16.3.4.1.

The sub-contractor has an obligation to comply with *all* of the requirements of the proviso if he considers that valuation of work executed should be in accordance with the rules in clause 16.3.4.

All too often sub-contractors fail either in timeous presentation of records for verification or in presentation of incomplete records, which they subsequently try and recover in final account negotiations. The sub-contract unfortunately makes no provision for action in the event of failure to abide by the requirements of the proviso, nor is there any established precedent, but it is suggested that, unless the sub-contractor complies with the requirements in full, the contractor has no liability to value, and subsequently make payment, in accordance with clause 16.3.4.

16.4

The whole of clause 16.4 on the face of it is very straightforward but, in the author's experience, is very rarely used by the quantity surveyor.

The intention is to reflect within the value of a variation the effect of that variation on other work in so far as it substantially changes the *conditions* under which that other work is executed. The sub-contractor has the right to have that other work treated as if it had been varied by an instruction and valued in accordance with the provisions of clause 16 but there are exceptions where amendment 7 applies:

(1) in respect of the expenditure of a provisional sum for defined work, other than to the extent that the work differs from the description for such work in the bill of quantities, or

(2) in respect of work for which an approximate quantity is included in the bills of quantities to such extent as the quantity is more or less than the quantity ascribed in those bills.

16.5   Addition to or deduction from sub-contract sum

Clause 16.5 should be read in conjunction with clause 3 and provides that any valuation under clause 16 is to be included in interim certificates and the sub-contractor has the right to be paid therefor.

17.3   Valuation rules

17.3   .1

The sub-contractor has both a right and an obligation to have work measured and valued in accordance with the rules, but only to the extent that the valuation relates to work *which can be properly valued by measurement.*

Measurement in this context, as confirmed in the later clause 17.3.2.1, relates to measurement following the same principles as those used in the preparation of the bills of quantities. If the particular work cannot be measured in accordance with those principles, then the provisions of clause 17.3.1 cannot apply and the sub-contractor must look to the other rules to determine which one should apply.

There follow sub-clauses .1, .2 and .3 which set out the rules to be adopted for valuation under the particular conditions appertaining, all of which contain both a right and an obligation to abide by those rules.

17.3   .1   .1

This on the face of it is clear but sub-contractors should note that all elements must be satisfied for the rule to apply. The work must be of similar character *and* executed under similar conditions *and* not significantly change the quantity. If any one element does not apply then this rule does not apply.

17.3 .1 .2

This is the next logical step after sub-clause .1, but again all required elements must be satisfied – it must be of similar character but *not* executed under similar conditions *and/or* significantly change the quantity. Two elements only need be satisfied here; if they are not then this rule cannot be used.

17.3 .1 .3

This is the next logical step after sub-clause 2 – the whole of clause 17.3 is logical in the way that it is set out such that once all parameters are satisfied, that is the rule to apply. In this case only one element has to be satisfied, i.e. that the work is not of similar character.

Provided the opening to clause 17.3.1 is satisfied, i.e. that the work can be properly valued by measurement, one of the sub-clauses .1, .2 or .3 must contain the rule that is to apply to the valuation of the variation.

17.3 .2

Clause 17.3.1 gives a right to and places an obligation on sub-contractors to use and abide by those rules. The three sub-clauses to clause 17.3.2, which apply to all valuations under clauses 17.3.1, are no different; both parties to the sub-contract must abide by these rules.

17.3 .2 .1

Reference was made to this clause in 17.3.1 above. Unless the work can be properly valued by measurement, the provisions of clause 17.3.1 cannot apply and this clause sets out that those measurements shall be in accordance with the same principles as those governing the preparation of the bills of quantities comprised in the sub-contract documents. Neither party to the sub-contract can depart from those measurement rules.

17.3 .2 .2

It should be noted that percentage additions or lump sum adjustments apply to all valuations under clause 17.3.1, *including work valued at fair rates and prices under clause 17.3.1.3,* and are not restricted to those items valued in accordance with clauses 17.3.1.1 and 17.3.1.2, i.e. at rates set out in the bills of quantities or pro rata thereto.

17.3   .2   .3

This is often a difficult rule to apply since generally sub-contractors do not, and indeed are often not given the opportunity to, price preliminaries separately.

Where the preliminaries are priced separately, this clause requires those preliminaries to be included in the valuation of work; it further requires that in respect of variations or in regard to the expenditure of provisional sums included in the sub-contract documents, except with regard to expenditure of a provisional sum for defined work (see clause 11.10.5.1 under Amendment 7 for comments on 'defined'), the preliminaries are to be adjusted to take account of those matters and the sub-contractor has a right to due and proper consideration of preliminaries in respect of all work carried out.

Where preliminaries are not priced separately, compliance with clauses 17.3.1.1 and 17.3.1.2 should not cause a problem since the bill rates will be inflated by the amount of preliminaries built into the original bid and, whilst this may not be totally satisfactory, it does give the sub-contractor a general recovery of pre-liminaries across all works which should be sufficient to cover costs.

Valuation under clause 17.3.1.3 can prove difficult where there is no breakdown of the preliminaries in the original bid, but, irrespective of a provided breakdown or otherwise, sub-contractors have the right to have preliminaries considered in any valuation of their work.

17.3   .3

Note the wording. This clause relates *only* to work which *cannot be valued by measurement.* It must be repeated that there is a logical progression to the way in which the rules in clause 17.3 are set out, but often sub-contractors will turn to this clause immediately as a way of ensuring that all costs are covered. *This is wrong.* Sub-contractors and contractors must follow the logical progression within clause 17.3 and use the rule which first applies to the circumstances surrounding the valuation. The basic question is, 'Is it measurable?'. If the answer is yes, clause 17.3.1 will apply; if the answer is no, clause 17.3.3 will apply.

17.3 .3 .1, .2

There is a tendency for 'specialist trades' to believe that they fall within the province of clause 17.3.3.2, no matter what their trade is. Reference to the wording of clause 17.3.3.2 and footnote [k] clearly shows that this is not true; the only trades falling within the rule in clause 17.3.3.2 are those in which the Royal Institution of Chartered Surveyors and the appropriate body representing the employers in that trade have agreed and issued a definition of prime cost of daywork. Footnote [k] confirms those bodies, and thereby the trades they represent, as being the Electrical Contractors Association, the Electrical Contractors Association of Scotland and the Heating and Ventilating Contractors Association. All other 'specialist trades' are bound by the rules in clause 17.3.3.1.

The sub-contractor has an obligation to comply with *all* of the requirements of the proviso if he considers that valuation of work executed should be in accordance with the rules in clause 17.3.3.

All too often sub-contractors fail either in timeous presentation of records for verification or in presentation of incomplete records, which they subsequently try and recover in final account negotiations. The sub-contract unfortunately makes no provision for action in the event of failure to abide by the requirements of the proviso, nor is there any established precedent, but it is suggested that, unless the sub-contractor complies with the requirements in full, the contractor has no liability to value, and subsequently make payment, in accordance with clause 17.3.3.

17.3 .4

As with clause 16.4, in the author's experience clause 17.3.4 is very rarely used by the quantity surveyor and is subject to an overall proviso (see later).

Again the intention is to reflect within the value of work executed the effect of directions on other work, except for directions in respect of the expenditure of a provisional sum for defined work other than to the extent that the work differs from the description for such work in the bill of quantities or in respect of work not executed, in so far as they substantially change the *conditions* under which that other work, which is not the subject of a direction, is executed, and the sub-contractor has the right to have that other work treated as if it had been varied by a direction and valued in accordance with the provisions of clause 17.3.1.2.

17.3 .5

This is the final rule and gives the sub-contractor the right to a fair valuation in the event that the valuation does not relate to or cannot reasonably be effected by the circumstances listed.

The overall proviso to clause 17.3 is of great importance and the sub-contractor is obliged to abide by the terms stated therein, i.e. that no allowance is to be made in a clause 17.3 Valuation for delay, disruption or other disturbance costs which are reimbursable under any other provision in the sub-contract, i.e. clause 13 which is the provision under which the sub-contractor's claims for direct loss and/or expense would be reimbursed.

18 Bill of quantities – standard method of measurement

18.1 Preparation of bills of quantities – errors in preparation etc.

18.1 .1

Note particularly the words in lines 1 and 2 commencing after 'unless'. Bland statements such as 'The Sub-

Contractor is deemed to abide by the measurement methods included within the Sub-Contract Documents' are in direct conflict with the requirements of clause 18.1 and cannot be enforced. Specific departures which draw the sub-contractor's attention to those departures such as 'Notwithstanding the requirements of clause ? of SMM7, this has been measured by etc.' do comply with the requirements of this clause and must be accepted.

18.1 .2

This clause gives the sub-contractor the right to have any departures, errors or omissions corrected; no written direction is necessary to have the relevant item treated as a variation, but the clause does not give the sub-contractor an automatic right to re-measurement in the event that clause 15.1 applies.

Clause 18.1.2 clarifies requirements in respect of defined provisional sums and states in line 5 that 'where the description of a provisional sum for defined work does not provide the information required by General Rule 10.3 in the *Standard Method of Measurement* the correction shall be made by correcting the description so that it does provide such information; any such correction under clause 18.1.2 shall be treated as if it were a Variation required by a direction of the Contractor'. Any corrections will be valued in accordance with clause 16.3 or 17.3 as appropriate and will require adjustment for any preliminaries etc. associated with the item, but any extensions of time and/or loss and expense issues arising from the expenditure of the provisional sum (due allowance for programming, planning and pricing preliminaries is deemed to have been made only for 'defined' provisional sums (General Rule 10.4)) come under the provisions of clauses 11 and 13.

18.1 .3

21.7

No additional rights or obligations are embodied; the amendments give effect to the adjustment of any 'Approximate Quantities' arising from the use of SMM7.

Part 3

Domestic Sub-Contract DOM/2

Introduction

The Domestic Sub-Contract DOM/2, published by the Building Employers Confederation, is intended for use where the form of main contract is the JCT Standard Form 'With Contractor's Design' (1981) and is approved by the Building Employers Confederation, the Specialist Engineering Contractors Group, the Federation of Associations of Specialists and Sub-Contractors and the Federation of Building Specialist Contractors. It is *not* suitable for use with any other version of the JCT Standard Form.

The Sub-Contract DOM/2 is based on the Sub-Contract DOM/1, amended by way of a schedule attached to the DOM/2 Articles of Agreement, and, as it is so similar to DOM/1, only relevant differences, additional rights and/or obligations or any relevant additional comments are considered in the following pages of this chapter.

A total of seven amendments have been issued since the first edition in 1981, all now being incorporated as standard. Amendment 5 to Sub-Contract DOM/2 (published in September 1989) is, however, only for use where bills of quantity included in the sub-contract numbered documents have been prepared in accordance with SMM7.

As with the Sub-Contract DOM/1, DOM/2 comprises two documents: the Articles of Agreement and the Sub-Contract Conditions. This Part will begin by considering the articles and will move on to the conditions later.

The Articles of Agreement

First recital

The contractor desires to have work designed and executed as referred to in the appendix; the footnote confirms that if the sub-contractor is not designing the sub-contract works the words 'designed and' are to be deleted, in which case the sub-contractor has no design responsibility and his sole responsibility is to execute the sub-contract works to details provided by the contractor and in accordance with the sub-contract.

Third recital

The sub-contractor has had reasonable opportunity of inspecting all of the provisions of the main contract.

Note here that the sub-contractor has a right to inspect 'all of the provisions of the Main Contract, or a copy thereof, except the detailed prices of the Contractor included in the Employer's Requirements, the Contractor's Proposals or the Contract Sum Analysis' and the sub-contractor must be given a reasonable opportunity of carrying out that inspection.

Note that DOM/2 introduces references to the employer's requirements, the contractor's proposals and the contract sum analysis as opposed to the schedules and bills of quantities in DOM/1, but the sub-contractor should still ensure that he does inspect *all* of the documents before entering into contract, for reasons already given in this book.

Article 1

1.3 The sub-contract conditions referred to in the schedule to the articles of agreement and in the list attached thereto are deemed to be incorporated. Provision is made within this article to incorporate any amendments which are to apply. This is slightly different to DOM/1 because DOM/2 has its own set of amendments which are additional to those in DOM/1, but both sets have to be incorporated into the Sub-Contract DOM/2. Article 1.3 refers, therefore, to the Sub-Contract DOM/2 and any amendments thereto, whilst the opening article to the schedule to Sub-Contract DOM/2 refers to the Sub-Contract Conditions DOM/1 and thus both sets of amendments are incorporated into the Sub-Contract DOM/2.

Appendix to DOM/2

Part 1 section B provides for completion of details of the main contract appendix 1 and entries therein, and is similar in all respects to part 1, section B of DOM/1.

Part 1 section C, which is unique to DOM/2, provides for completion of details of the main contract appendix 2 and entries therein in respect of the method of payment alternatives, alternative A covering stage payments and alternative B covering periodic payments.

Part 1 section D, which is also unique to DOM/2, provides for completion of details of the main contract appendix 3 and entries therein in respect of documents comprising the employer's requirements, the contractor's proposals and the contract sum analysis.

Part 1 section E provides for completion of details in respect of obligations or restrictions imposed by the employer and not covered by the main contract conditions, any employer's requirements affecting the order of the works and the location and type of access, and is the same in all respects to part 1, section C of DOM/1.

As with DOM/1, all of the above should contain nothing new and should merely be confirmation of what the sub-contractor should already know.

The Conditions

Some of the required changes to DOM/1, such as 'Delete DOM/1 and Insert DOM/2' or 'Delete Architect and Insert Employer', are considered to be minor only, do not change the rights or obligations of the sub-contractor and are not, therefore, considered relevant to the purpose of this book.

1.3

Clause 1.3 contains merely a list of definitions with no rights or obligations on either party to the sub-contract, but sub-contractors should note particularly that under the Sub-Contract DOM/2 there is no architect or quantity surveyor, their definitions having been deleted in the Schedule to DOM/2, and any rights or obligations in DOM/1 arising from the actions of architect or quantity surveyor do not apply to DOM/2.

2.1   .3

Clause 2.1.3 was introduced with amendment 2 and now constitutes, in part, an anomaly in that clauses 2.1.1 and 2.1.2 referred to in clause 2.1.3.1 no longer exist in DOM/1.

Clause 2.1.3.1 is, therefore, superfluous, but clause 2.1.3.2 applies and obliges both parties to the sub-contract not to divulge or use except for the purposes of the sub-contract any confidential information of the other party save as provided in clause 2.1.3.

4.1

Again there is an anomaly in clause 4.1 in that the alterations detailed in amendment 2 have not been subsequently amended to take account of further amendments to DOM/1. Thus clause 4.1.4 was re-numbered as clause 4.1.5 and a new clause 4.1.4 introduced, but subsequently DOM/1 introduced a new clause 4.1.4 and DOM/2 has not been amended to reflect this change. Applying amendments 2 to 7 to the Sub-Contract DOM/1 incorporating amendments 1 to 3, 5, 6, 8 and 9 results in two clauses numbered 4.1.4. For the purposes of this narrative it makes sense to assume that clauses 4.1.1 to .4 remain as written in DOM/1, clause 4.1.5 of DOM/1 is re-numbered to clause 4.1.6, and the clause 4.1.4 introduced as a result of amendment 2 to DOM/2 is re-numbered to clause 4.1.5.

4.1   .2, .3, .4

These three sub-clauses all contain the word 'shall', which places an obligation on the sub-contractor to ensure that all materials, goods and workmanship comply with the standards.

As with DOM/1, failure to comply with these requirements can have disastrous effects on the financial outcome of a project, since the sub-contract confers upon the contractor considerable powers in the event of breaches of these requirements, and sub-contractors should again ensure, by way of quality management systems or the like, that everything, as far as possible, complies with the documents. Unlike DOM/1, however, there are no situations under DOM/2 where approval of quality or standards can be a matter for the opinion of the architect, since there is no architect under the main contract, and the responsibility for materials, goods and workmanship is totally that of the sub-contractor.

4.1   .5

This is a totally new clause and obliges the sub-contractor to provide samples, before carrying out relevant work and/or ordering materials, of the standard of workmanship or the quality of the goods or materials which the sub-contractor intends to provide, *but only to the extent that is specifically referred to in the sub-contract documents,* i.e. if samples are not specifically requested within the documentation, the sub-contractor is under no obligation to provide the samples unless directed by instruction under clause 4.2 of the sub-contract, in which case the direction would be a variation to be valued in accordance with the valuation rules laid down in clause 16.3 or clause 17.3 as appropriate.

4.2 .1
Amendment 2 introduces a proviso that the contractor may not order a variation which is, or makes necessary, an alteration or modification to the design of the sub-contract works without the consent of the sub-contractor. Clearly this proviso can only apply where the sub-contractor has assumed responsibility for the design of the sub-contract works, its intent being that any decisions regarding the integrity of any design work carried out by the sub-contractor should remain with the sub-contractor, e.g. if the sub-contractor has designed the ventilation system he will be responsible for the performance of that system to the specification contained within the documentation and the contractor will not have any right to order a change, say, in the type of fan, possibly for a cheaper model, without the consent of the sub-contractor.

4.3 .2
The Sub-Contract DOM/1 contains two provisos to the operation of clause 4.3 but DOM/2 contains only one. The first proviso, relating to materials or goods or workmanship where approval of the quality or standards is a matter for the opinion of the architect, does not apply to DOM/2 (see clauses 4.1.2 and 4.1.3 above), but that relating to instructions issued under clause 8.4 of the main contract is retained exactly as in the Sub-Contract DOM/1.

4.3 .3, .4
The original clause 4.3.3 is deleted as a result of there being no architect under the main contract and subsequent clause 4.3.4 and 4.3.5 re-numbered to 4.3.3 and 4.3.4 respectively. As a result of the deletion of the original clause 4.3.3 the sub-contractor's rights and obligations regarding non-complying work are lost. The rights and obligations in the re-numbered clauses are unaffected by the cosmetic amendments thereto.

5.1 .1
The schedule deletes references to clauses 7 and 16 of the main contract, and the sub-contractor, therefore, has *no* obligation to observe, perform and comply with those main contract conditions regarding 'Levels and setting out

of the Works' and 'Materials and goods unfixed or off-site'.

5.3   Sub-contractor's liability for design

This is a new clause and sets out the sub-contractor's liability for design *but only to the extent that the sub-contractor has designed the sub-contract works. If the sub-contractor has not designed the sub-contract works he has no liability whatsoever for design.*

5.3   .1

The sub-contractor is obliged to accept like liability in respect of any defect or insufficiency in design as would an architect or other appropriate professional designer holding himself out as competent to take on such design.

The sub-contractor should be aware that liability under this clause is not absolute – he has no liability for example for 'fitness for purpose' – but most employers and/or contractors seek to amend this clause in some way to increase the sub-contractor's liability, many introducing 'fitness for purpose' requirements. Sub-contractors should be aware that any extension to their standard liability may affect their professional indemnity insurance and this should be checked with their insurance company before accepting any amendments to the standard wording of the clause. Any extra premium required by the insurance company, assuming they are prepared to take on the extra risk, should be added to the tender price, if possible as an optional extra so that employers and contractors will become aware of the cost of the additional liability.

5.3   .2

If the design work under the sub-contract is in connection with a dwelling or dwellings, the sub-contractor's liability under clause 5.3.1 includes liability under the Defective Premises Act 1972 and the sub-contractor is obliged to do all that is necessary for a document or documents to be issued for the purpose of Section 2(1) of the Act, but only where Section 2(1) of the act is included in the employer's requirements as referred to in appendix 1 of the main contract (part 1, section B, of the appendix to DOM/2).

5.3 .3

The sub-contractor has the right to the benefit of any operation of clause 2.5.3 of the main contract conditions (loss of profit or other consequential loss arising from liability under clause 2.5.1 of the main contract – clause 5.3.1 of the Sub-Contract DOM/1) to the extent that any contribution sought by the contractor can relate only to that liability not excluded by the operation of clause 2.5.3 of the main contract conditions.

This clause relates *only* to loss of profit or other consequential loss arising from a defect or insufficiency in design as referred to in clause 5.3.1 and no further.

8A.1

The insurance required to be taken out by the contractor is extended to cover the cost of design work, but again attention is drawn to comments contained within the DOM/1 section of this book that the contractor's liability and the sub-contractor's right are limited to *loss or damage caused by the specified perils only*. Footnote [e] is still applicable and sub-contractors should satisfy themselves that their own insurance is adequate to cover any additional design work arising out of loss or damage which is not as a result of the specified perils or, in respect of any design work sub-sub-let, that the sub-sub-contractor's insurance is adequate to cover all liabilities required.

11.10

As with the Sub-Contract DOM/1, clause 11.10 does not give any rights to or place any obligations on the sub-contractor; it is merely a list of events which may give rise to an extension of time under clause 11 but, again, only provided that all of the requirements of clause 11 have been complied with by the sub-contractor. The events are generally all as the Sub-Contract DOM/1, but event 11.10.7 (delay on the part of nominated sub-contractors or of nominated suppliers) is not used, event 11.10.13 is renumbered as 11.10.15 and a new event 11.10.13, relating to changes in statutory requirements after the date of tender, introduced.

13.3   .3

As with the Sub-Contract DOM/1, these sub-clauses merely list those matters which can give rise to the recovery of direct loss and/or expense referred to in clause 13.1 and contain no additional rights or obligations over and above those listed in clauses 13.1 and 13.2. However in the Sub-Contract DOM/2, clause 13.3.3, relating to any discrepancy in or divergence between documents, is deleted. In the event of discrepancy or divergence clause 2.4 of the main contract conditions confirms that the contractor's proposals prevail; any divergence from this would constitute a change in the employer's requirements in accordance with clause 12 of the main contract and any loss and expense arising therefrom is covered under relevant matter clause 13.3.7.

13.3   .8

Clause 13.3.8 is a new relevant matter unique to the Sub-Contract DOM/2 and covers the situation where loss and/or expense has arisen as a result of delay in receipt of any permission or approval for the purposes of development control requirements.

14.3   Liability of sub-contractor for defects in sub-contract works

This clause obliges the sub-contractor to take responsibility for any defects etc. which occur due to failure by the sub-contractor to comply with his obligations under the sub-contract, or due to frost occurring before the date of practical completion of the sub-contract works.

This obligation is greater than the comparable clause under the Sub-Contract DOM/1, which only requires the sub-contractor to take responsibility for faulty workmanship and/or materials or due to frost occurring before practical completion of the sub-contract works. As with DOM/1, there are some points worthy of expansion:

(1)   Clause 17 of the main contract covers partial possession by the employer and clearly the sub-contractor should be aware of any such possession since it affects his insurance requirements under the sub-contract

and triggers the start of the defects liability period specified in the main contract for that part.

(2) The 'without prejudice' statement prior to the liability statement in this clause is to ensure that the sub-contractor's liability for defects is no greater than that of the contractor under the main contract. Without this statement the sub-contractor would have a never-ending liability for defects, whereas with it his liability is limited to the defects liability period specified in the main contract.

(3) Defects due to frost occurring before practical completion are, in the author's experience, very rare but note the words in the clause: frost occurring before the date of practical completion of the *sub-contract works*. It is possible for frost to affect the sub-contract works after they are practically complete and it is, therefore, essential, in order to protect the sub-contractor from any future liability, to comply with the requirements of clause 14.1 and notify the contractor of the date when, in the sub-contractor's opinion, the sub-contract works are practically completed.

14.4 Employer's instructions – clauses 16.2 and 16.3 of the main contract conditions

As with clauses 17.2 and 17.3 of the main contract conditions (JCT 80) in the Sub-Contract DOM/1, clauses 16.2 and 16.3 of the main contract conditions (JCT 81) give the employer the right to make an appropriate deduction in respect of any defects, shrinkages or other faults not made good and clause 14.4 of the sub-contract steps down that right and obliges the sub-contractor to accept his share of the deduction.

Such a course of action is extremely rare as employers will generally require defects etc. to be made good, but there can be instances where access for the making good can be virtually impossible or the act of making good would disrupt operations to such an extent that the costs to the employer would be prohibitive. Under such circumstances an appropriate deduction would be the only sensible solution and hence this clause.

Clauses 16.2 and 16.3 of the main contract conditions do not appear to limit the employer's rights in any way, but do not, in the author's view, give the employer the right to make a deduction for all defects – such an action would be unreasonable – nor do they specify what an 'appropriate deduction' is or how this is to be calculated; this can only be assessed based on the likely costs of making good and would be subject to agreement by all parties to the contract or sub-contract as appropriate.

21.4   Payment for off-site goods and materials

.1   .3

The sub-contractor has a right to payment for off-site goods and materials where so specified in alternative A of appendix 2 of the main contract conditions, otherwise payment is purely at the discretion of the employer as referred to in alternative B of appendix 2 of the main contract conditions, and the sub-contractor is obliged to observe all relevant conditions set out in the main contract which have to be fulfilled before the employer is empowered to include such an amount.

The various comments against the relevant clause in the Sub-Contract DOM/1 covering payment for materials off-site are equally relevant to the Sub-Contract DOM/2.

21.7   .1
Final adjustment of sub-contract sum

Where clause 15.1 applies (lump sum contract), the sub-contractor is obliged to send to the contractor, *not later than 2 months after practical completion of the sub-contract works* (see clause 14), all documents necessary for the purpose of the adjustment of the sub-contract sum.

Note that the period for submission of documents is half that in the Sub-Contract DOM/1 because, unlike JCT 80 contracts where production of the final account is the responsibility of the quantity surveyor named in the contract, JCT 81 requires the contractor to submit the final account to the employer within 3 months of practical completion of the works. The various comments against the relevant clause in the Sub-Contract DOM/1 are equally relevant to the Sub-Contract DOM/2.

21.7 .3

This is a totally new clause introduced in amendment 2 and provides for procedures in the event that the sub-contractor fails to submit to the contractor the documents referred to in clause 21.7.1 within the 2 months specified therein.

21.7 .3 .1, .2

If the sub-contractor fails to provide documents as aforesaid, the contractor *may* give notice in writing that if the said documents are not submitted within 2 months from the date of the written notice, the contractor may finally adjust the sub-contract sum in accordance with clause 21.7.3.2.

Note that the contractor is *not* obliged to give notice in writing but, in the event that he chooses to adopt the procedures in clause 21.7.3, the sub-contractor has the right to receive the notice as specified.

21.7 .3 .3

If the sub-contractor does not dispute anything in the contractor's sub-contract sum in writing, giving grounds for so disputing, within 1 month of the date of its being sent to the sub-contractor, the sub-contractor is obliged to accept the contractor's sub-contract sum as conclusive as to the balance due between the parties.

Sub-contractors should note that the grounds given for disputing the sum will have to be in detail; a simple statement of non-agreement with the figures produced would fail to satisfy the requirements of the clause. Clearly it is not in the sub-contractor's interests that clause 21.7.3 should be invoked and sub-contractors should, therefore, ensure that all documents required by clause 21.7.1 are submitted within the time scale stated.

21.8 .1

Computation of ascertained final sub-contract sum

Where clause 15.2 applies (subject to complete re-measurement and valuation), the sub-contractor is obliged to send to the contractor, *not later than 2 months after practical completion of the sub-contract works* (see clause 14), all documents necessary for the purpose of computing the ascertained final sub-contract sum.

Again sub-contractors should note that the period for submission of documents is half that in the Sub-Contract DOM/1 and again previous comments regarding clause 21.7.1 and the Sub-Contract DOM/1 are equally applicable to clause 21.8.1 and the Sub-Contract DOM/2.

21.8 .3

As with clause 21.7.3, this is a totally new clause introduced in Amendment 2 and provides for procedures in the event that the sub-contractor fails to submit to the contractor the documents referred to in clause 21.8.1 within the 2 months specified therein.

21.8 .3 .1, .2

If the sub-contractor fails to provide documents as aforesaid, the contractor *may* give notice in writing that if the said documents are not submitted within 2 months from the date of the written notice, the contractor may finally adjust the sub-contract sum in accordance with clause 21.8.3.2.

Note that the contractor is *not* obliged to give notice in writing but, in the event that he chooses to adopt the procedures in clause 21.8.3, the sub-contractor has the right to receive the notice as specified.

21.8 .3 .3

If the sub-contractor does not dispute anything in the contractor's sub-contract sum in writing, giving grounds for so disputing, within 1 month of the date of its being sent to the sub-contractor, the sub-contractor is obliged to accept the contractor's sub-contract sum as conclusive as to the balance due between the parties.

This obligation is exactly the same as clause 21.7.3.3 and sub-contractors should again note that the grounds given for disputing the sum will have to be in detail; a simple statement of non-agreement with the figures produced would fail to satisfy the requirements of the clause. Again it is not in the sub-contractor's interests that clause 21.8.3 should be invoked and sub-contractors should, therefore, ensure that all documents required by clause 21.8.1 are submitted within the time scale stated.

21.9   Amount due in final payment

Date of final payment

21.9   .1, .2

The revised wording to clause 21.9.1 introduced by amendment 2 gives effect to the 'Contractor's Ascertained Final Sub-Contract Sum' arising out of the operation of clause 21.7.3 or clause 21.8.3 as appropriate, and the revised wording to clause 21.9.2 gives effect to when the final payment becomes due, having regard to the procedures for submitting and agreeing the final account under the main contract clauses 30.5.1, 30.5.5 and 30.5.6. Amendment 4 amends the time scales for when the final payment becomes due and when it is made, but neither of the amendments 2 or 4 substantially affect the sub-contractor's rights as described within the Sub-Contract DOM/1. However, as the time scales change slightly, the sub-contractors rights are reproduced below so that they can be more clearly understood.

The sub-contractor has the right to be paid the amount calculated in accordance with clause 21.7 or 21.8 as appropriate, less only discount and previous payments as stated, and the final payment becomes due not later than 14 days after the date that the contractor's final account and final statement, by the operation of clause 30.5.5 of the main contract conditions, or the employer's final account and final statement, by the operation of clause 30.5.6 of the main contract conditions, become conclusive as to the balance due between employer and contractor. Before the date that final payment becomes due the sub-contractor has the right to be notified by the contractor in writing by registered post or recorded delivery of the amount of the final payment to be made and, finally, the sub-contractor has the right to be paid the final payment within 21 days of the date that it becomes due.

This is all very clear and straightforward except that sub-contractors generally are not aware of the date of submission of the final account and the final statement under clause 30.5.1 of the main contract conditions and the contractor is not under any obligation to tell the sub-contractor of the date of their submission. Nor will the sub-contractor generally be aware if the contractor has failed to comply with the requirements of clause 30.5.1 of the main contract conditions leading to the possible invoking of clause 30.5.6; the employer is not obliged to produce the final account and final statement. As with clause 21.7.3 or clause 21.8.3 of the Sub-Contract DOM/2, it is an option only and, as long as it is not done, the employer is not under any obligation to pay any money, as the contractor under the sub-contract is equally not obliged to pay.

The sub-contractor's clear and straightforward right suddenly becomes clouded and, as in clause 21.5 of DOM/1, monitoring becomes necessary, using as guides the date of practical completion of the works, the defects liability period, the date of the issue of the notice of completion of making good defects under clause 16.4 and the requirements of clause 30.5 of the main contract conditions. Clause 30.5.5 (and 30.5.8 if clause 30.5.6 is invoked) of the main contract confirms that the final statement becomes conclusive 1 month after whichever of the listed events last occurs and the date that final payment becomes due and payable under the Sub-Contract DOM/2 is linked to when the final statement becomes so conclusive. It is, therefore, possible to calculate approximately this likely date when the final payment should become due, assuming a 12 months defects liability period, as shown in the following table.

Event	1 (months)	2 (months)	3 (months)
Practical completion of the works	0	0	0
Defects liability period in main contract	12	12	
Make good defects/Issue of notice of completion of defects		3	
Issue final account and final statement			3
1 month period	1	1	1
Total period from practical completion of the works	13	16	4

Each column in the table above represents the total period from practical completion of the works to the issue of the final certificate in respect of each of the events listed in clause 30.5.5:

1. the end of the defects liability period
2. the issue of the Notice of Completion of Defects,
3. the submission of the final account and the final statement required by clause 30.5.1 of the main contract.

It can be seen that the earliest date for the issue of the final certificate on a contract with a 12 months defects liability period is 13 months from practical completion of the works, but column 2 is more likely to apply to most contracts. Subcontractors should, however, carefully check the defects liability period since 6 months for building work and 12 months for mechanical and electrical installations and landscaping only are quite common and, whilst the period for mechanical and electrical etc. will determine the date for issue of the final payment, a notice of completion of defects for building work will determine the date for final release of retention monies for that part (see clause 21.5). Should the contractor fail to comply with his obligations under clause 30.5.1 of the main contract conditions such that the employer invokes clause 30.5.6, this will add time but in practical terms only to column 3. Although the contract does not give any time constraints for the employer's production of the final account and final statement, it is in the employer's interests to know what his final financial liability is going to be and he is unlikely to leave production of the final account for too long. The time by which column 3 becomes extended, therefore, is unlikely to be greater than the overall period in column 2, although by no means impossible.

26.3

This is a totally new clause introduced by amendment 2.

If the sub-contractor wishes to sub-let the design for all or any portion of the sub-contract works he is obliged to obtain the written consent of the contractor before so doing and such consent is not to be unreasonably withheld, but the sub-contractor remains wholly responsible for the design in accordance with his obligations under clause 5.3.

37.3 .4

Non-adjustable elements only apply where the employer is a local authority.

38.2 .1

If the dispute or difference to be referred to arbitration raises issues which are substantially the same as or connected with issues raised in a related dispute between parties as listed, and if the related dispute has already been referred for determination to an arbitrator, the contractor and the sub-contractor have agreed and are obliged to refer the dispute or difference under the sub-contract to the arbitrator appointed to determine the related dispute, i.e. all references to nominated sub-contractors or nominated suppliers are removed.

Part 4

Further Developments

Introduction

Sub-contracts have to keep pace with current events as they arise from both changes in the law and changes in custom and practice. This chapter will look briefly at some issues which are likely to affect the use and development of the Sub-Contracts DOM/1 and DOM/2 in the future:

(1)   further amendments to DOM/1 and DOM/2
(2)   the Arbitration Act 1996
(3)   small claims procedures
(4)   the Housing Grants, Construction and Regeneration Act 1996.

Further Amendments to DOM/1 and DOM/2

The Sub-Contracts DOM/1 and DOM/2 incorporating amendments 1 to 9 inclusive are not totally compatible with the current versions of JCT 80 in use. The Joint Contracts Tribunal has issued amendments up to amendment 15 and amendment TC/94 for use with JCT 80, but the Building Employers Confederation has not yet issued consequential amendments to DOM/1 after amendment 9. Sub-contractors may, therefore, be working or proposing to work on contracts where their rights and obligations are different to the contractor's rights and obligations under the main contract and the JCT amendments will now be briefly considered to determine what loss of right or additional obligation or additional risk, if any, the sub-contractor is assuming as a result of the disparity.

Amendment 10 to JCT 80

Amendment 10, issued in March 1991, re-drafted the procedure for nomination of a sub-contractor and has no consequential effect upon the Sub-Contracts DOM/1 and DOM/2.

Amendment 11 to JCT 80

Amendment 11, issued in July 1992, consisted of eight items amending clauses concerning:

1. *Contractor's insurance – personal injury or death – injury or damage to property*
   The amendment requires the contractor's insurance to indemnify the employer in respect of claims for personal injury to or death of any person employed by the contractor. The equivalent clause in DOM/1, clause 7, in its present form requires no such indemnity, although the generality of clause 5.1 of DOM/1 could mean that sub-contractors should consider the main

contract amendment when negotiating conditions with their insurers.

2. *Determination by employer*
    The determination provisions of the main contract clause 27 have been re-drafted but the re-draft does not affect the rights or obligations of the sub-contractor under the provisions of clause 31 and clause 30.2 of the sub-contract.

3. *Determination by contractor*
    The main contract clause 28 has been similarly re-drafted but again the re-draft does not affect the rights or obligations of the sub-contractor under the provisions of clause 31 and clause 30.2 of the sub-contract.

4. *Determination by employer or contractor*
    As with the main contract clauses 27 and 28, the re-draft does not affect the rights or obligations of the sub-contractor under the provisions of clause 31 and clause 30.2 of the sub-contract.

5. *Outbreak of hostilities*
    Clause 32 is deleted and noted as 'Number not used'. The sub-contractor's specific obligation under the terms of clause 5.1 of the sub-contract is, therefore, now non-existent, but the main contract clause 28A, at 28A.1.1.5 and 28A.1.1.6, includes the new events 'hostilities involving the United Kingdom (whether war be declared or not)' and 'terrorist activity' as events which can give rise to determination of the employment of the contractor.

6. *War damage*
    The current statutory position in respect of war damage is explained in the guidance notes to amendment 11. Clauses 33.3 and 33.4 have no current application and clause 33.2 became inoperative with the decision to delete clause 32. Clause 33 is, therefore, deleted and noted as 'Number not used'. As a result, the sub-contractor's specific obligation under the terms of clause 5.1 of the sub-contract is, as with clause 32, now non-existent.

7. *Nominated sub-contractors*
    As the amendment was the re-draft of a clause concerning nominated sub-contractors there can be no consequential effect upon the sub-contracts DOM/1 and DOM/2.

8. *Various amendments consequential on the above*
    As can be seen from the above, there is very little consequential effect of amendment 11 upon the rights and

obligation of the sub-contractor under the sub-contract and for all practical purposes the amendment can be ignored. The effect of the various consequential amendments can similarly be ignored.

Amendment 11 also made a correction to the Without Quantities Private and Local Authorities versions of JCT 80 which sub-contractors can ignore as regards its effect upon the sub-contract.

Amendment 12 to JCT 80

Amendment 12, issued in July 1993, introduced a new clause 42 covering 'Performance Specified Work'. Practice note 25 defines performance specified work as 'comprising materials and components or assemblies of a kind or standard to satisfy design requirements given in the tender documents for the contract' and quotes as examples trussed rafters, precast concrete floor units or simple installations for lighting, heating or power. Clearly the very nature of this amendment is likely to require sub-contract involvement and, without some amendment to DOM/1, sub-contractors may well be asked to price performance specified work in bills of quantities without any consequential definition or contractual arrangement within the sub-contract.

The author is aware of one major contractor who seeks to embody the principles of amendment 12 (and amendments 11, 13 etc.) by 'deeming them to be incorporated with words deemed suitably amended until such time as the BEC produce their own amendments, following which the BEC amendments will be deemed to be incorporated'. Quite how such wording would be interpreted by the courts in the event of dispute is not known, but how an amendment which has not in itself been amended to suit the sub-contract or how something not yet written can be deemed to be incorporated into the sub-contract is in itself a recipe for dispute and conflict.

Sub-contractors faced with having to quote against performance specified work in bills of quantities should seek to have the sub-contract clearly and unequivocally amended to incorporate rights and obligations, and, more importantly, to clearly establish their liabilities, at a level adequate to cover what the performance specification expects. It is suggested that the incorporation of a document based upon amendment 12, re-written to suit the sub-contract (with the permission of the Joint Contracts Tribunal), would clearly and unequivocally establish such rights, obligations and liabilities. As an alternative it may be possible to amend the wording of clause 5.1 to introduce a specific obligation on the sub-contractor to observe, perform and comply with the provisions of the main contract conditions clause 42, which would then give the sub-contractor the benefit of the operation of clause 42 as a condition of the sub-contract, but this may cause problems as to how the necessary consequential amendments to the sub-contract would be incorporated.

The fact that the Building Employers Confederation have not yet been able to publish an amendment to DOM/1 to incorporate performance specified work 3 years after its introduction by the Joint Contracts Tribunal is an indication of the difficulty in writing a suitable document, but sub-contractors need adequate documentation and should insist upon same as a condition of their quotation.

The detailed amendments which would be necessary for such a document are beyond the scope of this book, but examination of the rights and obligations of the contractor under clause 42, and, therefore, the sub-contractor in a suitable contractual chain (references to the architect could be replaced by 'the architect through the contractor or the contractor' and references to the contractor would be replaced by 'the Sub-Contractor'), may give sub-contractors an insight into their rights, obligations and liabilities in respect of performance specified work:

Part 5: Performance Specified Work

Immediate attention is drawn to footnote [ee] which, in turn, draws attention to paragraphs 2.6 to 2.8 of practice note 25 for a description of work which is *not* to be treated as performance specified work:

1. Where it is intended simply to use manufacturer's standard components such as windows or kitchen fittings.
2. Where detailed design of other work would be dependent upon information provided in the contractor's statement or statements for performance specified work.
3. Where items would materially affect the appearance of

the building, or may result in changes in the design of other work, or would affect the use of the building such that it would be essential to examine and accept the contractor's proposals for the work before acceptance of the tender.

42.1 Meaning of performance specified work

42.2 Contractor's statement

Before carrying out any performance specified work, the contractor is obliged to provide the architect with a document or set of documents, the documents being referred to as the contractor's statement and complying with the requirements of clause 42.3.

42.3 Contents of contractor's statement

42.4 Time for contractor's statement

The contractor is obliged to provide his statement by the date specified in the clause, i.e. by the date given in the contract bills or by any reasonable date given in the instruction of the architect on the expenditure of a provisional sum for performance specified work or, if no such date is given, at a reasonable time before the contractor intends to carry out the performance specified work.

As far as the sub-contractor is concerned, the date for provision of the statement is likely to be included in the enquiry since the enquiry will either be for work specified in the contract bills or work for which an architect's instruction will have already been issued. It is unlikely, but by no means impossible, that a provisional sum for performance specified work will be included within sub-contract bills of quantities.

42.5 Architect's notice to amend contractor's statement

The architect may, within 14 days after receipt of the contractor's statement, issue a notice in writing requiring the contractor to amend his statement if, in the architect's opinion, the statement is deficient as specified in the clause and the contractor is obliged to so amend his statement such that in the architect's opinion, it ceases to be deficient.

The architect is not under any obligation to issue the notice, but the contractor is not under any obligation to amend his statement unless the notice is issued. The contractor is, however, obliged to provide the architect with a copy of any amended statement (amendment 14). Notwithstanding the issue or otherwise of a notice under this clause, the contractor remains responsible for any deficiency in his statement and for the work to which such statement refers.

42.6 Architect's notice of deficiency in contractor's statement

If the architect finds anything in the contractor's statement which appears to be a deficiency which would adversely affect the performance of the relevant work, he is obliged to issue a notice to the contractor specifying the discrepancy and the contractor, therefore, has the right to receive such notice.

The architect is under no obligation to find any deficiency; his only obligation is in the event that he does find a deficiency. As with clause 42.5, notwithstanding the issue or otherwise of a notice under this clause, the contractor remains responsible for the performance specified work.

42.7 Definition of provisional sum for performance specified work

42.8 Instructions of the architect on other provisional sums

The definition of a provisional sum for performance specified work contained in clause 42.7 confirms that such a provisional sum has to be provided in the contract bills. Clause 42.8 requires that no instruction of the architect on the expenditure of other provisional sums shall require performance specified work and the contractor is, therefore, under no obligation to carry out performance specified work other than that specified as such in the contract bills.

42.9 Preparation of contract bills

42.10 Provisional sum for performance specified work – errors or omissions in contract bills

If there is any error or omission in the information included in the contract bills in respect of a provisional sum for performance specified work, the contractor has both a right and obligation to correct such error or omission and any such correction is to be treated as if it were a variation required by an instruction of the architect.

The information required to be provided is specified in clause 42.7 and any required correction does not require a written instruction from the architect.

42.11 Variations in respect of performance specified work

The architect has the right to issue instructions requiring a variation to performance specified work and the contractor is obliged, per se, to comply with those instructions, but subject to the provisions of clause 42.12.

42.12 Agreement for additional performance specified work

The architect has no right to issue instructions requiring performance specified work additional to that envisaged in the contract unless the employer and the contractor otherwise agree. It is important to understand and appreciate that any agreement must be between the employer and the contractor; the architect is not empowered to enter into any such agreement and, if the contractor did enter into agreement with the architect, the employer would not then be bound to honour that agreement.

42.13 Analysis

The contractor is obliged to provide an analysis of the portion of the contract sum which relates to performance specified work where the contract bills do not provide such analysis, but only if required to do so by the architect and he is then further obliged to provide the analysis within 14 days.

42.14   Integration of performance specified work

The architect is obliged to give, and the contractor, therefore, has the right to receive, instructions necessary for the integration of performance specified work with the design of the works, and the contractor is obliged to comply with such instructions.

It is important that contractors understand that integration of performance specified work is the responsibility of the architect, and contractors should not undertake that responsibility either by instruction or by agreement otherwise than with the client, since such action would be a breach of the requirements of clause 42.12.

42.15   Compliance with architect's instructions – contractor's notice of injurious affection

If the contractor is of the opinion that any instruction of the architect in respect of performance specified work is likely to affect the performance of that work, he is obliged, within 7 days of receipt of the relevant instruction, to give notice to the architect in writing specifying the effect. Unless the architect amends his instruction to remove the effect, the contractor is not obliged to carry out the instruction without his (the contractor's) consent, although his consent is not to be unreasonably withheld or delayed.

Withholding of consent under this clause must be reasonable; a variation which cannot be accommodated other than by over-stressing the equipment being provided would be unreasonable and consent could reasonably be withheld, but a variation which can be accommodated by changing the equipment, notwithstanding the obligation to give notice as required by clause 42.15, could be given consent.

42.16   Delay by the contractor in providing the contractor's statement

Where the architect has not received the contractor's statement within the time period included within clause 42.4 or any amendment as a result of the operation of clause 42.5 and the failure of the contractor as aforesaid has affected the progress of the works, the contractor will lose his right to an extension of time and recovery of loss and expense to the extent of the delay so caused.

42.17   Performance specified work – contractor's obligation
42.17   .1

The contractor is obliged to exercise reasonable skill and care in the provision of performance specified work, but such obligation cannot be construed as to affect the contractor's obligations under the contract in respect of workmanship, materials or goods nor can anything in the contract operate as a guarantee of fitness for purpose of the performance specified work.

42.17   .2
The contractor's obligation under clause 42.17.1 is unaffected by any service which he has obtained from others in respect of the performance specified work, e.g. if there is an element of design work in achieving his obligations for the performance specified work and he sub-lets that design element, the contractor retains responsibility for the whole of the work notwithstanding that a portion was sub-let.

42.18   Nomination excluded

Sub-contractors will realise from the rights and obligations detailed above the importance of suitably amending the Sub-Contract DOM/1 to incorporate similar provisions. Incorporation of an equivalent clause 42, say as clause 39, is not, however, the end of the matter; other alterations need to be made to the text of DOM/1 to make the clause workable. The following are the author's suggestions:

clause 1.3:	to equate to revised main contract clause 1.3
clause 4.1.5.3	new clause to equate to new main contract clause 2.4.1
clause 4.1.5.4	new clause to equate to new main contract clause 2.4.2
clause 4.1.6	new clause to equate to new main contract clause 5.9
clause 4.1.7	new clause to equate to new main contract clause 8.1.4

clause 11.10.5	to equate to revised main contract clause 25.4.5.1
clause 11.10.16	new clause to equate to new main contract clause 25.4.15
clause 13.3.7	to equate to revised main contract clause 26.2.7
clause 14.1	to equate to revised main contract clause 17.1
clause 16.1 and	
clause 17.1	to equate to revised main contract clause 13.4.1.1
clause 16.3 and	
clause 17.3	to equate to revised main contract clause 13.5
clause 21.10.1.1	to equate to revised main contract clause 30.10

Any contractual amendments to incorporate the effects of main contract clause 42 need to be dealt with by experts and it is essential that, if faced with tendering for performance specified work without suitable contractual amendments being proposed by the enquirer, sub-contractors should approach their legal adviser on how to proceed further.

Finally in amendment 12 a new relevant event was introduced: 'the use or threat of terrorism and/or the activity of the relevant authorities in dealing with such use or threat'. A consequential amendment to DOM/1 should be made if sub-contract work is being carried out under the provisions of a main contract incorporating this amendment.

Amendment 13 to JCT 80

Amendment 13, issued in January 1994, introduced four items amending:

1. a re-draft of clause 13.2 to recognise a new clause 13A
2. a new clause 13A
3. consequential amendments arising from items 1 and 2
4. amendments to clauses 27.6.2.1 and 27.6.2.2.

The guidance notes to amendment 13 confirm that the amendment introduces a new method of valuing a variation by providing for the contractor to submit a quotation for any variation, the quotation to include for any direct loss and/or expense and for any change of time required for completion of the works and arising from the execution of the variation. The advantage of this method to the contractor (and, therefore, the sub-contractor in a suitable contractual chain) is that he will know, before carrying out the variation, the amount of the payment he will receive, including the amount of any disturbance (loss and/or expense), and the revised completion date. The operation of the quotation method is subject to being so required on an instruction of the architect and cannot be applied to any variation; the giving of quotations against instructions is not unknown but, unless the instruction states that a quotation is required, any quotation is of no effect and the valuation must be carried out in accordance with the valuation rules contained within the contract.

Clearly, where amendment 13 is incorporated into the main contract, a consequential amendment will need to be made to any sub-contract to keep the contractual chain intact and, as with amendment 12, the various rights and obligation under amendment 13 will now be considered, although references to nominated sub-contractors are ignored since clearly they cannot apply to sub-contractors appointed under the Sub-Contract DOM/1.

13A Variation instruction – contractor's quotation in compliance with the instruction

The re-draft of clause 13.2 (instructions requiring a variation) was necessary in order to introduce the new clause 13A as a contractual requirement. There is no directly equivalent clause within the Sub-Contract DOM/1; any architect's instruction under the main contract would arise in the sub-contract under a contractor's direction pursuant to clause 4.2. Logically, therefore, there needs to be an amendment within clause 4.2 to introduce the quotation requirement and maintain the contractual chain, but within DOM/1 there are two clauses covering valuation rules – clauses 16 and 17 – and there may well be a requirement for quotations under either lump sum or re-measurement sub-contracts. It will, therefore, be necessary to accommodate both situations by suitably amending clause 4.2.2 to make the provision of a quotation required by the architect binding on the sub-contractor and by introducing clauses equivalent to clause 13A to cover the required situations; it is suggested as clauses 16A and 17A.

13A Contractor to submit his quotation ('13A Quotation')

Clause 13A only applies where the contractor has not registered disagreement, pursuant to clause 13.2.3, with its application.

Clause 13.2.3 obliges the contractor, if he disagrees with the application of clause 13A, to write and so state within 7 days (or such other period as may be agreed) of receipt of the instruction referring to clause 13A.

The time scale under the main contract is very small, unless some other period is agreed, and, if sub-contractors are to have the same right of objection, the time scale within the sub-contract will have to be even smaller to the extent that it may well become unworkable if the disagreement under the main contract is to be registered within 7 days.

13A.1 .1

The instruction to which clause 13A is to apply is to provide sufficient information to enable the contractor to provide a quotation and, if the contractor considers that the information provided is not sufficient, he is obliged, not later than 7 days from receipt of the instruction, to request the architect to supply sufficient further information.

Attention is drawn to footnote [i.2] which confirms that the information provided should normally be in a similar format to that provided at tender stage and may be in the form of drawings and/or an addendum bill of quantities and/or a specification or otherwise. It does not, however, have to be in the same form as the main contract, i.e. with quantities where the form of main contract is the 'with quantities' version; any form which adequately communicates the requirements is acceptable.

13A.1 .2

The contractor is obliged to submit his quotation to the quantity surveyor not later than 21 days from receipt of the instruction or the receipt of sufficient further information to which clause 13A.1.1 refers, whichever is the later, and is to remain open for acceptance by the employer within 7 days from its receipt by the quantity surveyor.

Note that acceptance of the quotation is the prerogative of the employer, not the architect, and that any sub-contract will need to have extended time periods for acceptance in order for time periods under the main contract to operate effectively.

13A.1 .3
The contractor is obliged not to carry out the variation for which he has submitted his quotation until receipt by the contractor of the confirmed acceptance issued by the architect pursuant to clause 13A.3.2.

13A.2 Content of the contractor's 13A Quotation

The contractor is obliged to provide a quotation which separately comprises the elements listed in sub-clauses 13A.2.1 to .6 inclusive:

.1 the value of the required adjustment to the contract sum supported by all necessary calculations and including, where appropriate, allowances for any adjustment of preliminary items

.2 any adjustment to the time required for completion of the works

.3 the amount to be paid for any direct loss and/or expense in lieu of ascertainment of same under clause 26.1

.4 a fair and reasonable amount in respect of the cost of preparing the quotation

and, but only where specifically required by the originating instruction, indicative information in respect of:

.5 additional resources required, if any, to carry out the variation, and

.6 the method of carrying out the variation.

Each part of the quotation is required to contain sufficient supporting information to enable each constituent part to be properly evaluated by or on behalf of the employer. Sub-contractors will be aware of the advantages of having the price of variation work agreed prior to carrying out the said work and the requirement to provide information should not be taken lightly. Failure to provide sufficient information could lead to an instruction to carry out the work with the valuation to be carried out in accordance with the valuation rules or to an instruction not to carry out the work. There is no 'second bite of the cherry' as far as the giving of information is concerned – clause 13A contains no provision for the request of further and better particulars in connection with any quotation given and any quotation has to be either accepted or instructions issued to carry out or not to carry out the work (clause 13A.4).

13A.3 Acceptance of 13A Quotation – architect's confirmed acceptance

13A.3 .1
If the employer wishes to accept the quotation, he is obliged to notify the contractor in writing, and the contractor has the right, therefore, to receive such notice, not later than the last day of the period for acceptance referred to in clause 13A.1.2.

Note that the notification is required to be issued by the employer; any notification from any other source should, at the least, be queried before carrying out any work in view of the specific obligation required by this clause.

13A.3 .2
If the employer accepts a 13A Quotation, the architect is obliged to immediately confirm such acceptance by statements in writing as required by sub-clauses 13A.3.2.1 to .4, although clause 13A.3.2.4 will not apply to sub-contractors appointed under the Sub-Contract DOM/1 since it confirms the nominated sub-contract portion of any 13A Quotation.

Note that this confirmation is for the purpose of introducing the work into the contract and does not relieve the employer of his obligation under clause 13A.3.1.

13A.4 Contractor's 13A Quotation not accepted

If the employer does not accept the quotation by the expiry

of the period for acceptance stated in clause 13A.1.2, the architect is obliged to either instruct that the variation is to be carried out and valued in accordance with the rules of the contract or to instruct that the variation is not to be carried out and the contractor has the right, per se, to receive such instruction.

13A.5 Payment for a 13A Quotation

If a 13A Quotation is not accepted, the contractor has the right to be paid a fair and reasonable amount in respect of the cost of preparation of the quotation provided that the quotation has been prepared on a fair and reasonable basis.

In this context the non-acceptance of a quotation is not to be construed as evidence that the quotation has not been prepared on a fair and reasonable basis.

13A.6 Restriction on use of a 13A Quotation

If the architect has not issued a confirmed acceptance of a 13A Quotation, the employer and the contractor are obliged not to use that 13A Quotation for any purpose whatsoever.

The use of clause 13A is limited and the intention behind clause 13A.6 is that any 13A Quotation should not be used in subsequent negotiations either in respect of valuations under the valuation rules in the contract or any further requests for a 13A Quotation for the same or similar work.

13A.7 Number of days – clauses 13A.1.1 and/or 13A.1.2

13A.8 Variations to work for which a confirmed acceptance of a 13A Quotation has been issued – valuation

In the event of a variation to work for which a 13A Quotation has been accepted, the contractor has the right to have that variation valued on a fair and reasonable basis having regard to the content of the 13A Quotation and the valuation rules in the contract do not apply. The contractor has the further right to have included in the valuation any direct loss and/or expense incurred as a result of the regular progress of the works or any part thereof having been affected by compliance with the instruction requiring the variation.

Sub-contractors will again realise from the rights and obligations detailed above the importance of suitably amending the Sub-Contract DOM/1 to incorporate similar provisions. However, incorporation of an equivalent clause 13A, say as clauses 16A and 17A, is not, as with amendment 12, the end of the matter. Other alterations need to be made to the text of DOM/1 to make the clause workable; the following are the author's suggestions:

clause 1.3	to equate to revised main contract clause 1.3
clause 4.2.4	to equate to revised main contract clause 4.1.1 and introduce new clause to equate to main contract clause 4.1.1.2
clause 11.6	to equate to revised main contract clause 25.3.2
clause 11.7.2	to equate to revised main contract clause 25.3.3
clause 11.10.5.1	to equate to revised main contract clause 25.4.5.1
clause 12.1	to give effect to a revised completion date following the operation of a 'quotation in compliance with an instruction'
clause 13.3.7	to equate to revised main contract clause 26.2.7
clause 16.1 and clause 17.1	to equate to revised main contract clause 13.4.1.1
clause 16.5	to equate to revised main contract clause 13.7
clause 17.4	new clause to give effect to the operation of a 'quotation in compliance with an instruction' in a sub-contract subject to re-measurement
clause 18.1.1	to equate to revised main contract clause 2.2.2.1
clause 18.1.2	to equate to revised main contract clause 2.2.2.2
clause 21.7.2	to equate to revised main contract clause 30.6.2
clause 21.8.2	to give effect to the operation of a 'quotation in compliance with an instruction' in a sub-contract subject to re-measurement
clause 34.3	new clause to equate to new main contract clause 37.3.

As with amendment 12, any contractual amendments to incorporate the effects of main contract clause 13A need to be dealt with by experts and it is essential that sub-contractors should approach their legal adviser on how to proceed further if faced with tendering under the provisions of a main contract which incorporates amendment 13.

Finally in amendment 13, clause 27.6 (consequences of determination under clause 27.2 to 27.4) was re-drafted. This does not affect the sub-contractor's rights under clause 31 and clause 30.2 of the sub-contract, but the wording of clause 27.6.2.2 gives the employer the right to pay any supplier or sub-contractor for any materials or goods delivered or for any work carried out before or after the date of determination if the price has not already been discharged by the contractor. There are exceptions as to when the employer's right can be used but, when read with the case of *B Mullan & Sons (Contractors) Ltd* v. *John Ross and Malcolm London* (1996), this clause may have far-reaching consequences for sub-contractors.

Mullan were groundwork sub-contractors to McLaughlin & Harvey plc under a JCT 81 main contract. JCT 81, as with JCT 80, allows the employer to pay suppliers and sub-contractors once the main contractor's employment is terminated, but not if the main contractor has had a winding-up or liquidation order made. Under such circumstances there is a legal principle that everybody will bear the pain equally. Mullan asked the employer, Londonderry Port Commissioners, to pay direct (to Mullan) money owed by McLaughlin & Harvey, using, as the basis of its request, the argument that McLaughlin & Harvey could have no right or interest in any sum of money that the employer elected to pay direct to a sub-contractor. Under such circumstances, they argued, the Mullan money did not form a part of the main contractor's property and was not, therefore, available to the liquidator to pay out to all the creditors. The court decided that an expectation of payment was sufficient to give McLaughlin & Harvey an interest in the property, but the judge went on to draw a distinction between a company in receivership rather than being wound up and confirmed that, in receivership, employers could elect to make payments direct to sub-contractors. Any sub-contractor, therefore, faced with a main contractor in receivership could ask an employer to make direct payments before the company is wound up.

Amendment TC/94 to JCT 80

Amendment TC/94, issued in April 1994, introduced six items amending the insurance provisions of the main contract to specifically recognise terrorism. Whichever insurance clause applies (clause 22A or clause 22B or clause 22C), if terrorism cover becomes unavailable the employer has the option of bearing the risk himself or determining the employment of the contractor. Sub-contractors will be involved either by restoring damaged work, in which case they will be paid as if it were a variation, or, in the event of determination, the employment of the sub-contractor will in itself determine and the provisions of clauses 31 and clause 30.2 will apply. It is suggested that the definition of specified perils in the sub-contract should be amended to specifically recognise terrorism so that the sub-contractor may obtain the benefit of terrorism cover required under amendment TC/94.

Amendment 14 to JCT 80

Amendment 14, issued in March 1995, consisted of 14 items required to introduce the Construction (Design and Management) Regulations 1994 (CDM Regulations) as terms of the contract. The regulations place responsibilities on all parties to any construction contract, but the responsibilities under the main contract and the sub-contract are different and, therefore, the rights and obligations under the main contract will not be considered further. The CDM Regulations must, however, be incorporated into any sub-contract and the following are the author's suggested amendments, but again it must be stressed that expert help should be sought to properly amend any sub-contract:

clause 1.3   to equate to revised main contract clause 1.3

clause 4.1.4   to equate to revised main contract clause 8.1.3 – this will place an obligation on the sub-contractor to carry out his work in accordance with the health and safety plan developed by the principal contractor

clause 11.3.1   to give effect to any omission or default of the contractor's obligations as the principal con-tractor, where he is the principal contractor – any such omission or default will give the sub-contractor the right to an extension of time arising from such omission or default

clause 11.10.17   new clause to equate to new main contract clause

25.4.17 (the numbering assumes that a new clause 11.10.16 is introduced arising from main contract amendment 12) – the employer is required to ensure that the planning supervisor and the principal contractor, where he is not the contractor, carry out their duties under the CDM Regulations; this new clause gives the sub-contractor the right to an extension of time arising from the employer's compliance or non-compliance with the requirements under the main contract

clause 13.1
to give effect to any omission or default of the contractor's obligations as the principal contractor, where he is the principal contractor – any such omission or default will give the sub-contractor the right to recovery of loss and expense arising from such omission or default

clause 13.3.9
new clause to equate to new main contract clause 26.2.9 – this new clause gives the sub-contractor the right to recovery of loss and expense arising from the employer's compliance or non-compliance with his obligations under clause 6A.1 of the main contract

clause 14.1
to equate to revised main contract clause 17.1 – unless the sub-contractor has complied with the requirements of the new clause 20C, practical completion of the sub-contract works cannot be confirmed

clause 20C
new clause to equate to main contract clause 6A – equivalent clauses to clauses 6A.1 and 6A.2 will not be necessary, but additional clauses will be required to cover the interchange of information between the contractor and the sub-contractor to comply with the regulations. The sub-contractor will be obliged to provide such information to the contractor as the contractor considers reasonably necessary to enable the health and safety plan to be developed and monitored, together with any requirements of the principal contractor to the extent necessary to comply with the regulations

clause 29.1.5
new clause to equate to new main contract clause

27.2.1.5 – this will give the contractor the right to determine the employment of the sub-contractor under the sub-contract if the sub-contractor fails to comply with the requirements of the regulations

clause 30.1.1.4 new clause to equate to new main contract clause 28.2.1.4 – this will give the sub-contractor the right to determine his employment under the sub-contract if the contractor fails to comply with the requirements of the regulations.

Amendment 15 to JCT 80

Amendment 15, issued in July 1995, followed a judgment of the Court of Appeal in *Crown Estate Commissioners* v. *John Mowlem & Co Ltd* (1994) concerning the conclusiveness of a final certificate. The judgment provided an interpretation of the effect of a final certificate issued under JCT 80 which was much wider than intended by the Joint Contracts Tribunal in its original drafting. The judgment also gave wider effect to the certificate of practical completion and the certificate of making good defects. Amendment 15 revises the text of clauses 1.4 and 30.9.1.1 to negate the wider effects which could be interpreted from the original wording.

Clause 21.10.1.1 of DOM/1 contains the same wording as clause 30.9.1.1 of JCT 80 and logically, therefore, the effect of the final payment under the sub-contract would be subject to the same interpretation as the final certificate under the main contract. It therefore needs amending in a similar manner to clause 30.9.1.1 if the wider interpretation of the effect of the final payment is to be negated, although sub-contractors may not want such an interpretation to be narrowed.

It may well be that the Building Employers Confederation is so far behind with its amendments to the Sub-Contracts DOM/1 and DOM/2 because the future of those sub-contracts is uncertain. The Joint Contracts Tribunal, for example, has expressed an intention of writing a form of sub-contract for use with domestic sub-contracts and all sub-contracts will have to be at least examined, if not re-written, now that the Housing Grants, Construction and Regeneration Act 1996 has received royal assent. Even so those new or amended forms of sub-contract are unlikely to figure in contracts being prepared at present or in the immediate future and the Sub-Contracts DOM/1 and DOM/2 are likely to be in use for several years yet.

The Arbitration Act 1996

The Arbitration Act 1996, which supersedes the 1950 and 1979 Arbitration Acts, received royal assent in June 1996. At the time of going to press, an operative date was still to be announced. The new 1996 Act covers both domestic and international arbitration and contains many new provisions that were not previously to be found either in legislation or in common law.

Detailed discussion regarding all of the provisions of the new Act is beyond the scope of this book, but reference is included here to make sub-contractors aware that, once the Act is operative, recourse to arbitration under clause 38 of the Sub-Contract DOM/1 will be pursuant to the Arbitration Act 1996.

The object of arbitration, as the Act now defines it, is 'to obtain the fair resolution of disputes by an impartial tribunal without necessary delay or expense'. Each party to the dispute has to be given a reasonable opportunity of putting its case and dealing with that of its opponent and the parties are free to agree how their disputes are to be resolved, subject only to such safeguards as are necessary in the public interest. The arbitrator is required to act fairly and impartially as between the parties to the dispute and the parties are required to 'adopt procedures suitable to the circumstances of the particular case, avoiding unnecessary delay or expense', and to 'do all things necessary for the proper and expeditious conduct' of the proceedings.

The 1996 Act gives the arbitrator greater power (in some cases greater even than the courts) to determine how proceedings are to be conducted as well as evidential matters. He can, for example, determine what questions can be raised and answered in the proceedings. He will also have the power to decide whether and to what extent the hearing should itself take the initiative in ascertaining the facts and the law. Does this mean that the sub-contractor will lose his right of appeal in clause 38.7 of DOM/1? This is not to say that the arbitrator will knowingly interfere or will not act fairly and impartially, but changes to long established procedures may unwittingly lead to awards being made on the basis of incomplete facts.

The 1996 Act also gives the arbitrator power to direct that the recoverable costs of an arbitration should be limited to a specific amount. Thus the parties would be free to spend whatever they like, but in the knowledge that any recovery they make at the end of the day would be limited to the sum fixed by the arbitrator. Arbitrators may award simple or compound interest – even the courts, except in extremely rare circumstances, have no power to award compound interest.

For years before the publication of the act, there was debate as to whether arbitrators were immune from liability for anything done or omitted to be done in discharge of their function as arbitrators. The 1996 Act now gives that immunity.

The Act introduces a number of radical new concepts and procedures, and it has yet to be seen how successful arbitrators will be in putting these into practice. Having said that, the House of Lords was in favour of the bill with supportive comments coming from Lord Lester, Lord Donaldson and Lord Roskill and, with support from such eminent members of the legal profession, the Act must be given every chance to succeed. But sub-contractors must remember that anyone can act as an arbitrator and, in view of the new powers bestowed on the arbitrator by the Act, the choice of arbitrator becomes particularly important. Arbitrators will in future have to take the lead. No longer will they be able to sit back and wait for one party to ask them to do something, and they will have to be more imaginative and proactive in their conduct.

Whatever happens as a result of the new Act, the sub-contractors' right to have disputes resolved by arbitration will be unaffected but there will undoubtedly be changes to procedures.

Small Claims Procedures

The Sub-Contract DOM/1 provides arbitration as the only route to resolving disputes irrespective of the amount involved and many sub-contractors write off money because it would cost more in lawyers' fees to argue the claim than it is all worth. Arbitration is very expensive and risky, but a recent change in the small claims procedure in the county court may give sub-contractors an avenue for prosecuting smaller value claims with little cost and risk.

The small claims procedure in the county court was formerly limited to not more than £1,000, but that limit has been raised to £3,000 and sub-contractors who have a claim for an amount below this figure can use the small claims procedure to have the dispute resolved. An added bonus is that, if sub-contractors prosecute the claim themselves and lose, they do not have to pay the other side's costs. So, if a sub-contractor has a dispute over say £2,500 with a main contractor and the main contractor turns up with an army of lawyers, the main contractor has to pay for his lawyers, win or lose. The idea behind the small claims procedure is to dissuade parties from using lawyers; they can be used if so desired, but the lawyers' costs will almost certainly not be borne by the loser and the main contractor who receives a summons from a sub-contractor using this procedure will have to bear this in mind when deciding how and if to defend the action. For small value claims, therefore, it is well worth sub-contractors considering this procedure.

The Housing Grants, Construction and Regeneration Act 1996

The Housing Grants, Construction and Regeneration Act received royal assent in July 1996 but at the time of going to press no operative date had been announced and the final version of the Scheme for Construction Contracts was yet to be published. It seems that the Act will come into operation in spring 1997. The Scheme will operate if individual contracts do not make provision for the requirements of the Act.

Part II of the Act deals with construction contracts and was drafted in response to a number of proposals from the Latham Report on the construction industry in 1994. Not all of Sir Michael Latham's proposals were embodied in the new Act but this has addressed adjudication, payment, 'pay-when-paid' provisions, set-off and suspension for non-payment.

The impact of the Housing Grants, Construction and Regeneration Act 1996 on DOM/1 and DOM/2 is very limited, the Act being restricted to resolution of contentious issues, i.e. adjudication, payment and contra charges. The clauses affected are:

(a) *Adjudication:* The idea of adjudication to resolve disputes prior to instituting arbitration proceedings is a totally new concept. At present in DOM/1 and DOM/2, adjudication is only available under clause 24 to resolve contractors' claims not agreed by the sub-contractor. It is an interim measure only prior to arbitration and notice for arbitration has to be given at the same time as institution of the procedure.

(b) *Payment:* As written, clause 21 of DOM/1 and DOM/2 does not contain 'pay-when-paid' provisions, but many contractors amend the standard wording to include such provisions. Those amendments will be against the requirements of the Act and, therefore, unenforceable in the future.

(c) *Set-off:* The current set-off provisions in clause 23 comply with the requirements of the Act.

(d) *Suspension for non-payment:* Such a right is already embodied within clause 21.6 of DOM/1 and DOM/2.

Adjudication

Any party to a construction contract will have the right to refer a dispute to an adjudicator although it is left to the parties to the contract to agree their own procedure for adjudication. Any procedure must, however:

(a) be rapid
(b) impose a duty on the adjudicator to act impartially,
(c) enable the adjudicator to take the initiative in ascertaining the facts and the law
(d) give the adjudicator immunity from legal proceedings.

The adjudicator's decision must be binding until the dispute is finally determined by agreement, arbitration or litigation, though parties may agree to accept it as final.

If a contract does not provide for such a procedure, the relevant provisions of the Scheme for Construction Contracts (yet to be fully drafted) will apply but sub-contractors will realise that the procedural requirements leave much to be agreed between the parties, for example should the adjudicator or just an appointing body be named in advance?

The current adjudication procedure in clause 24 of the Sub-Contract DOM/1 could form the basis of a procedure to comply with the requirements of the Act if it was widened to cover any dispute and the time scales tightened.

Payment

Sub-contractors have suffered, particularly in recent years, from delays in payments and one of the proposals of the Latham Report was to put an end to 'pay-when-paid' clauses. Under the Act, pay-when-paid clauses, by which payers withhold payments pending receipt of payment from third parties, will be ineffective, except where a third party is insolvent. The Act includes in its list of insolvencies companies who have an administrative receiver or receiver and manager appointed and, under these circumstances, sub-contractors' attention is drawn to comments made previously under amendment 13 regarding the case of *B. Mullan & Sons (Contractors) Ltd* v. *John Ross and Malcolm London* (1996).

Contracts must provide an adequate mechanism for determining the amount and timing of payments. Contractors (and sub-contractors) may insist on interim payments, for contracts with a duration longer than 45 days, with a 'final date for payment' specified for each payment, although the parties are free to agree how long the period is to be between the date on which a sum becomes due and the final date for payment. This could mean that payments would be made, say, 60 days after the date they become due and comply with the requirements of the Act; would this be better than pay-when-paid?

A payer must give notice, within no more than 5 days from when a sum should fall due under a contract, of how much he intends to pay. If he then changes his mind, he must issue a further notice (of withholding) within the time allowed. Parties may suspend work if payment is withheld without the appropriate notices being served.

The Scheme for Construction Contracts, when issued, will determine payment terms and periods for notice to withhold payment if they are not specified by contract.

Set-off

Set-off is not allowed unless an effective written notice specifying the intent to withhold payment has been given. To be effective the notice must specify the amount proposed to be withheld and the ground or grounds for withholding payment; where there is more than one ground, each ground has to be specified and the amount attributable to it. If there is a dispute as to set-off, this, as with other disputes, will have to be referred to adjudication.

The current set-off rules contained within clause 23 of DOM/1 comply with the basic requirements of the Act, but go further. Clause 23 only allows set-off for amounts not agreed in respect of a contractor's claim for loss and/or expense and/or damage arising from a breach of, or failure to observe the provisions of, the sub-contract by the sub-contractor (clause 23.2.1) and requires any set-off amount to have been quantified in detail and with reasonable accuracy (clause 23.2.2). The Act does not prohibit set-off between different contracts and does not require set-off to be quantified. It may be possible, therefore, for unscrupulous contractors to give an effective notice without quantification for short term gain or to delay payment as a result.

Suspension for non-payment

Where a sum due is not paid in full by the final date for payment and no effective notice to withhold payment has been given, the party to whom payment is due has the right to suspend further execution of work under the contract but only after giving the offending party at least 7 days' notice of intention to suspend performance stating the ground or grounds on which it is intended to so suspend. This right is already contained within clause 21.6 of the Sub-Contracts DOM/1 and DOM/2.

Part 5

Flow Charts

Flow Charts

Contracts generally contain logical progressions dependent upon the interaction of the rights and obligations of each of the parties to the contract and, provided both parties abide by their individual rights and/or obligations, there is no reason why any contract should not reach a logical conclusion to the benefit of both parties. It is, however, sometimes difficult to fully appreciate and understand from the written word the logical progression of those rights and obligations and this lack of clear understanding can lead to breaches and dispute.

The flow charts on the following pages have been drawn to highlight the more common areas of dispute surrounding the Sub-Contracts DOM/1 and DOM/2 and cover:

(1) Contractor's directions under clause 4.
(2) Commencement and progress of the sub-contract works under clause 11, drawn with one completion date for simplicity but equally applicable to sections of the sub-contract works if required.
(3) Loss and expense and contractor's claims under clause 13.
(4) Valuation rules under clause 16.
(5) Valuation rules under clause 17.

Contractor's directions – Clause 4

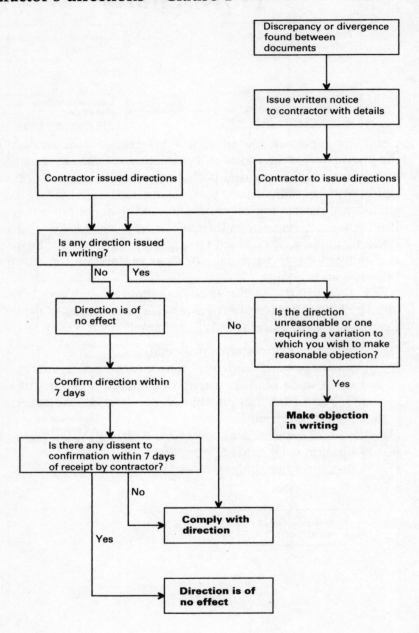

Contractor's directions – Clause 4 (cont.)

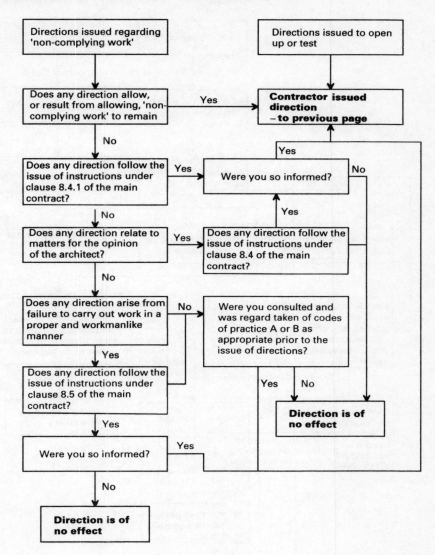

Commencement and progress – Clause 11

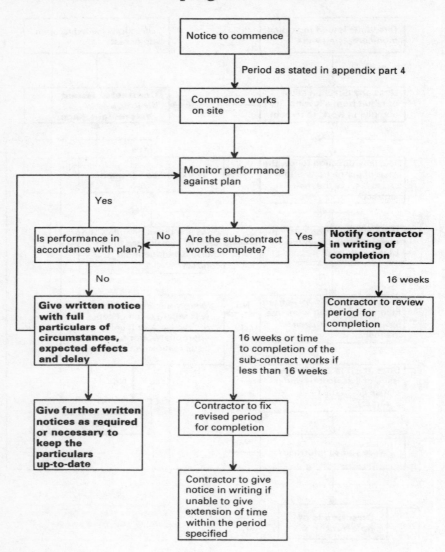

Loss and/or expense and contractor's claims – Clause 13

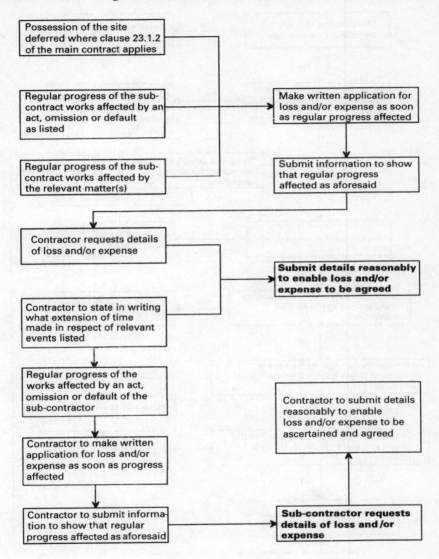

Valuation rules – Clause 16

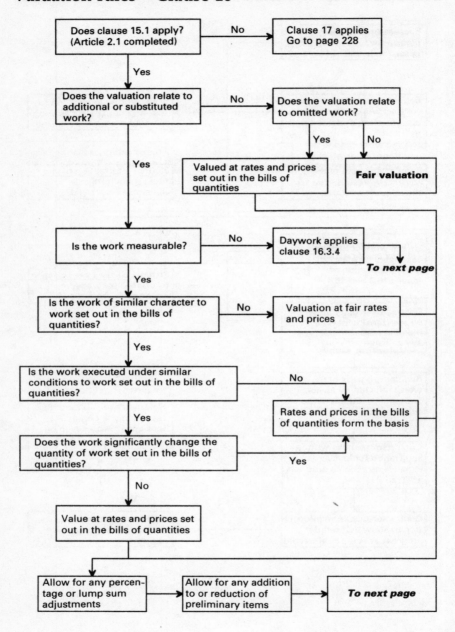

Does clause 15.1 apply? (Article 2.1 completed)	No →	**Clause 17 applies** Go to page 228

↓ Yes

Does the valuation relate to additional or substituted work?	No →	**Does the valuation relate** to omitted work?

Yes ↓ (additional)   Yes ↓   No ↓

Valued at rates and prices set out in the bills of quantities   **Fair valuation**

Is the work measurable?	No →	**Daywork applies** clause 16.3.4

To next page

↓ Yes

Is the work of similar character to work set out in the bills of quantities?	No →	**Valuation at fair rates** and prices

↓ Yes

Is the work executed under similar conditions to work set out in the bills of quantities?	No →

Rates and prices in the bills of quantities form the basis

↓ Yes   ↑ Yes

Does the work significantly change the quantity of work set out in the bills of quantities?	

↓ No

Value at rates and prices set out in the bills of quantities

Allow for any percen- tage or lump sum adjustments	→ **Allow for any addition** to or reduction of preliminary items	→ *To next page*

Valuation rules – Clause 16 (cont.)

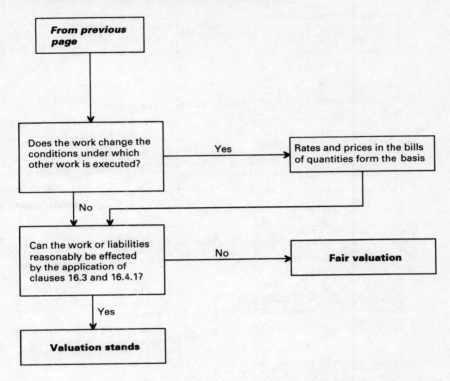

From previous page

↓

Does the work change the conditions under which other work is executed? —— Yes —→ Rates and prices in the bills of quantities form the basis

No ↓

Can the work or liabilities reasonably be effected by the application of clauses 16.3 and 16.4.1? —— No —→ **Fair valuation**

Yes ↓

Valuation stands

Valuation rules – Clause 17

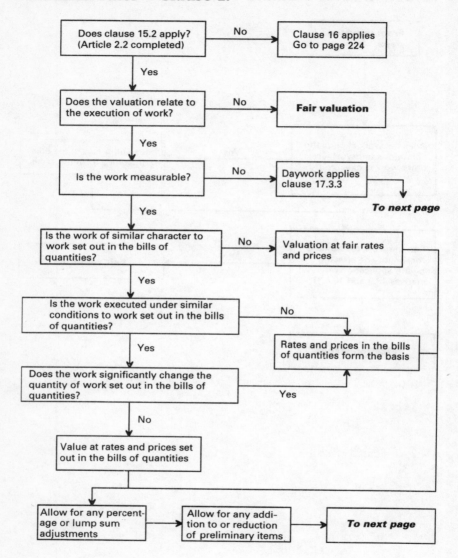

Does clause 15.2 apply? (Article 2.2 completed) → **No** → Clause 16 applies Go to page 224

↓ **Yes**

Does the valuation relate to the execution of work? → **No** → **Fair valuation**

↓ **Yes**

Is the work measurable? → **No** → Daywork applies clause 17.3.3 → *To next page*

↓ **Yes**

Is the work of similar character to work set out in the bills of quantities? → **No** → Valuation at fair rates and prices

↓ **Yes**

Is the work executed under similar conditions to work set out in the bills of quantities? → **No** →

↓ **Yes**

Does the work significantly change the quantity of work set out in the bills of quantities? → **Yes** → Rates and prices in the bills of quantities form the basis

↓ **No**

Value at rates and prices set out in the bills of quantities

↓

Allow for any percentage or lump sum adjustments → Allow for any addition to or reduction of preliminary items → *To next page*

Valuation rules – Clause 17 (cont.)

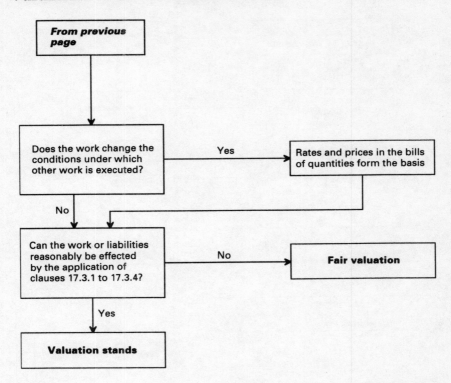

From previous page

Does the work change the conditions under which other work is executed?

Yes → Rates and prices in the bills of quantities form the basis

No

Can the work or liabilities reasonably be effected by the application of clauses 17.3.1 to 17.3.4?

No → **Fair valuation**

Yes

Valuation stands

Indexes

Clause Index

Obligations Index

Rights Index

Subject Index

Adjudication
action in set-off, 110
not to be confused with arbitration, 111

Arbitration
appeal to the High Court, 141
the Arbitration Act 1996, 211
dispute or difference, 140
JCT Arbitration Rules, 141
related dispute, 140
related dispute; DOM/2, 184
written notice, 140

Attendance
adequate provision at sub-contract
    placement, 116
clearance of rubbish, 116
facilities to be provided free of charge, 115
fitness, condition or suitability of
    scaffolding, 118
hoisting facilities, 115
items detailed in the appendix part 9, 116
keeping scaffolding erected, 117
level and cost of supply of temporary
    services, 117
materials on site, 115
scaffolding and scaffold boards for work
    11 feet high or under, 115
scaffolding and scaffold boards for work
    over 11 feet high, 115
security, 115
storage of materials, 115
temporary services as contractor supplied
    attendances, 117
temporary services to workshops, sheds
    or other temporary buildings, 115
use of erected scaffolding of the
    contractor or sub-contractor, 117
workshops, etc. of sub-contractor, 116

Bills of quantities
defined provisional sums, 163
departures, errors or omissions, 78, 163
measurement methods, 78, 163
preparation, 77, 162

Contract
ability to carry out work, 4
acceptance, 3, 4
basic requirements, 4, 5
by action, 4
capable of honouring, 4

capacity to contract, 3
consideration, 3, 5
counter-offer, 5
duress, 3, 5
formal signing, 4
the Housing Grants, Construction and
    Regeneration Act 1996, 215
illegal act, 3
legal binding sub-contract, 4
legal obligation, 3
offer, 3, 4
promise, 3
sub-contracts, 4
what is a contract?, 3

Contractor's claims
acceptance of the details, 61
ascertained and agreed, 61
cause and effect, 60
contractor's obligations, 60
default of the sub-contractor, 59
details of loss and/or expense, 61
direct loss and/or expense, 59
information, 60
provisos, 60
requests by the sub-contractor, 61
requests from the sub-contractor, 60
right to recover, 59
within a reasonable time, 60
written application, 59, 60

Defects
code of practice A, 143
code of practice B, 144
deduction if not made good, 66
deduction if not made good; DOM/2,
    177
frost occurring before practical
    completion, 65
liability of sub-contractor, 65
liability of sub-contractor; DOM/2, 176
materials or workmanship not in
    accordance with the sub-contract, 65
partial possession, 65

Determination
amount of any direct loss and/or
    damage, 120
benefit of any agreement, 119
by sub-contractor – right to be paid, 122
by sub-contractor – rights and duties of
    contractor and sub-contractor, 122

printed text of clauses 11.2 to.10, 129, 136, 140

recovery of increases applicable to sub-sub-contracts, 127, 134

refunds and premiums, 125, 132

reimbursement of fares etc., 131

relevant wage fixing body, 131

representations on the value of work to which the formula adjustment applies, 138

sub-contractor in default over completion, 129, 136

sub-contractor's profit, 128, 136

sub-let works – payment to or allowance by the sub-contractor, 127, 134

tender rates increased or decreased, 124, 131

tender rates increased or decreased; materials, 126

tender type ceases to be payable, 124, 131

tender type ceases to be payable; materials, 126

unforeseen and unrecoverable increase, 124

use of sub-contract/works contract formula rules, 137

valuations for the purpose of calculating formula adjustments, 137

wage rates etc. increased or decreased by promulgation, 130

work sub-let to sub-sub-contractors, 126, 133

workpeople engaged upon or in connection with the sub-contract works, 125, 130, 132

written notice by sub-contractor, 127, 134

Insurance

additional insurance, 30, 32, 36, 41

amount of insurance cover, 31

benefits of the joint names policy, 30

costs of restoration, replacement or repair, 33, 37, 42

default by contractor, 47

default by sub-contractor, 46

discovery of any loss or damage, 35, 39, 44

documentary evidence, 46

domestic sub-contractors, 30, 32, 37, 41

Employer's Liability (Compulsory Insurance) Act 1969, 31

excess provisions of joint names policy, 32

extended to cover the cost of design work; DOM/2, 175

fully, finally and properly incorporated, 34, 39, 43

loss or damage after practical completion, 36, 40, 45

negligence, breach of statutory duty, etc. of the contractor, 34, 38, 43

off-site materials and goods, 101

other risks, 30

partial possession, 65

payment for restoration etc. of work done, 36, 40, 45

personal injury or death, 29

practical completion of sub-contract works, 63

proof by contractor, 47

protect work, 33, 37, 41

restoration of work lost or damaged, 34, 38, 43

restore damage, 35, 40, 44

specified perils, 29, 33, 37, 38, 43

specified perils; DOM/2, 175

sub-sub-contractors, 31, 33, 37, 41

third party liability – clause 8C, 42

third party or public liability policy, 30

written confirmation of agreement of practical completion, 64

JCT 80

amendment 10 – nomination procedure, 189

amendment 11 – eight various amendments, 189

amendment 12 – performance specified work, 191

amendment 13 – quotation for any variation, 198

amendment 14 – Construction (Design and Management) Regulations 1994, 206

amendment 15 – effect of the final payment, 208

amendment TC/94 – terrorism cover, 205

Loss and expense

act, omission or default of the contractor, 104

agreement of, 58

approximate quantity in the bills of quantities, 152

ascertainment and agreement, 62

bad estimating or other loss reasons, 59

cause and effect, 57

contractor's request, 58

default of the contractor, 56, 104

243